Essential Chemistry

Acids & Bases

3rd edition

STERLING
Education

Customer Satisfaction Guarantee

Your feedback is important because we strive to provide the highest quality educational materials. Email us comments or suggestions.

info@sterling–prep.com

We reply to emails – check your spam folder

3 2 1

ISBN-13: 979-8-8855706-5-7

Sterling Education materials are available at quantity discounts.

Contact info@sterling–prep.com

Sterling Education
6 Liberty Square #11
Boston, MA 02109

Published by Sterling Education

Printed in the U.S.A.

STERLING
Education

From the foundations of chemical reactions to the complex mechanisms of atomic particles, *Essential Chemistry Self-Teaching Guides* are a comprehensive compendium of clearly explained texts to learn and master these multifaceted chemistry topics.

These guides provide a detailed review of the fundamental mechanisms of chemical and physical processes at the atomic level. Develop a better understanding of the electronic structure of elements, principles of chemical bonding, phases of matter, types and mechanisms of chemical reactions, and principles of solutions and acid-base equilibria. Understand rate processes in chemical reactions, empirical and molecular formulas, enthalpy, entropy, oxidation number, the laws of thermodynamics, and electrochemistry. Apply the theoretical knowledge by working through the practice questions and step-by-step solutions.

Created by highly qualified chemistry instructors, researchers, and education specialists, these books empower readers by helping them increase their understanding of general chemistry.

We sincerely hope that these guides are valuable for your learning.

220525gdx

Essential Chemistry Self-Teaching Guides

Electronic Structure & Periodic Table

Chemical Bonding

States of Matter & Phase Equilibria

Stoichiometry

Solution Chemistry

Chemical Kinetics & Equilibrium

Acids & Bases

Chemical Thermodynamics

Electrochemistry

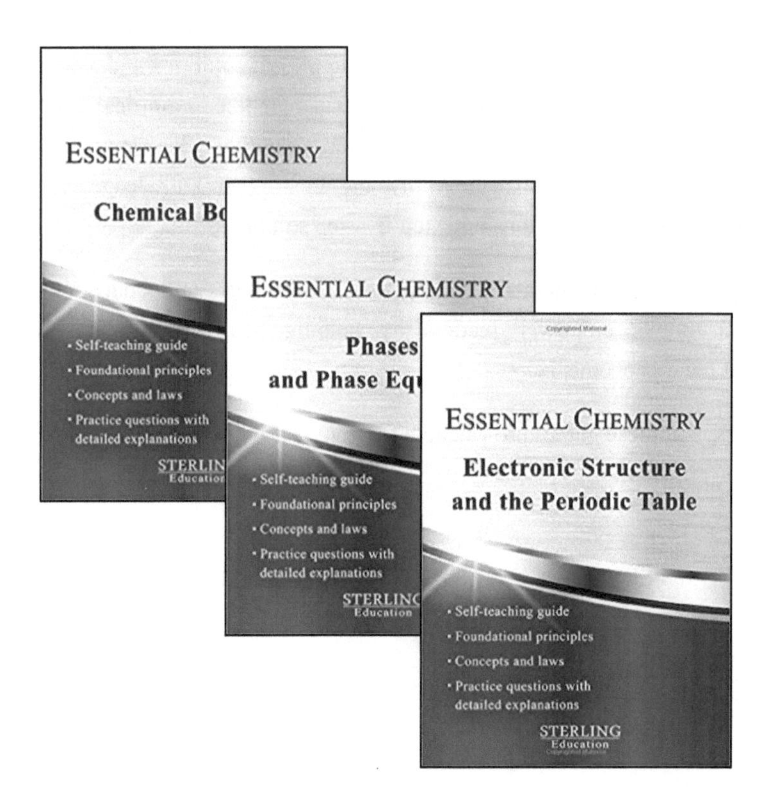

Essential Physics Self-Teaching Guides

Kinematics and Dynamics

Equilibrium and Momentum

Force, Motion, Gravitation

Work and Energy

Fluids and Solids

Waves and Periodic Motion

Light and Optics

Sound

Electrostatics and Electromagnetism

Electric Circuits

Heat and Thermodynamics

Atomic and Nuclear Structure

Visit our Amazon store

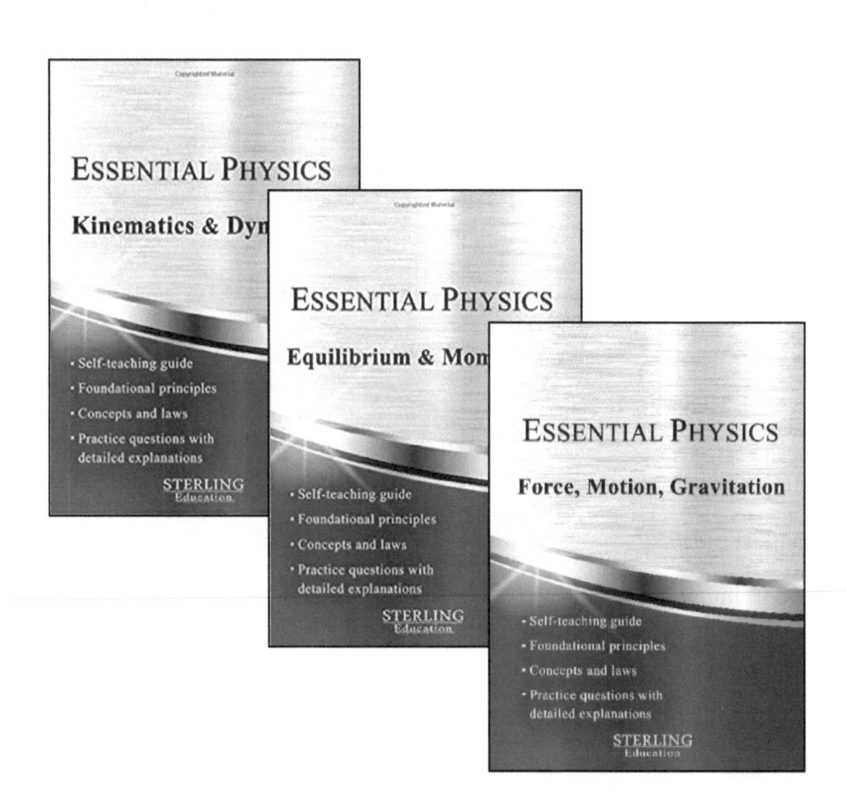

Essential Biology Self-Teaching Guides

Eukaryotic Cell: Structure and Function

Enzymes and Cellular Metabolism

DNA, Protein Synthesis, Gene Expression

Specialized Eukaryotic Cells

Genetics

Nervous System

Endocrine System

Circulatory System

Respiratory System

Lymphatic and Immune System

Digestive System

Excretory System

Skeletal System

Muscle System

Integumentary System

Reproductive System

Development

Microbiology

Plants

Photosynthesis

Evolution, Classification, Diversity

Ecosystems and Biosphere

Population and Community Ecology

Visit our Amazon store

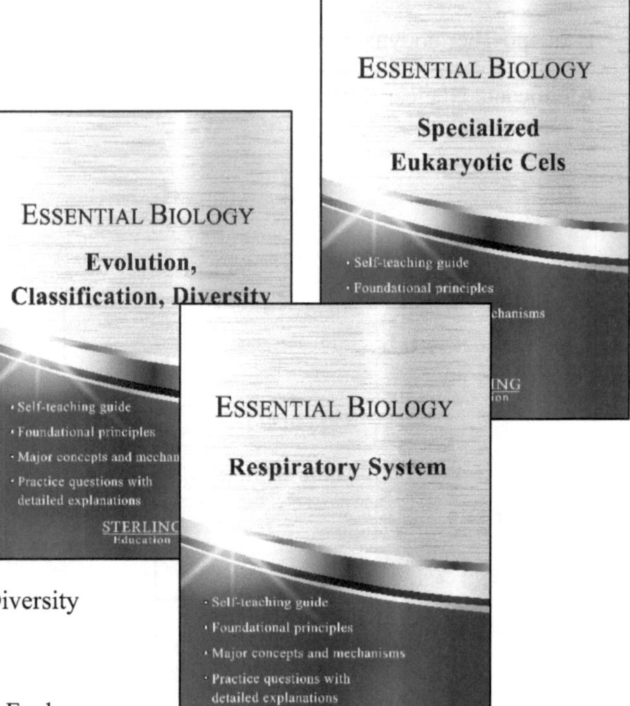

Everything You Always Wanted to Know About…

Chemistry

Physics

Cell and Molecular Biology

Organismal Biology

American History

American Law

American Government and Politics

Comparative Government and Politics

World History

European History

Psychology

Environmental Science

Human Geography

Visit our Amazon store

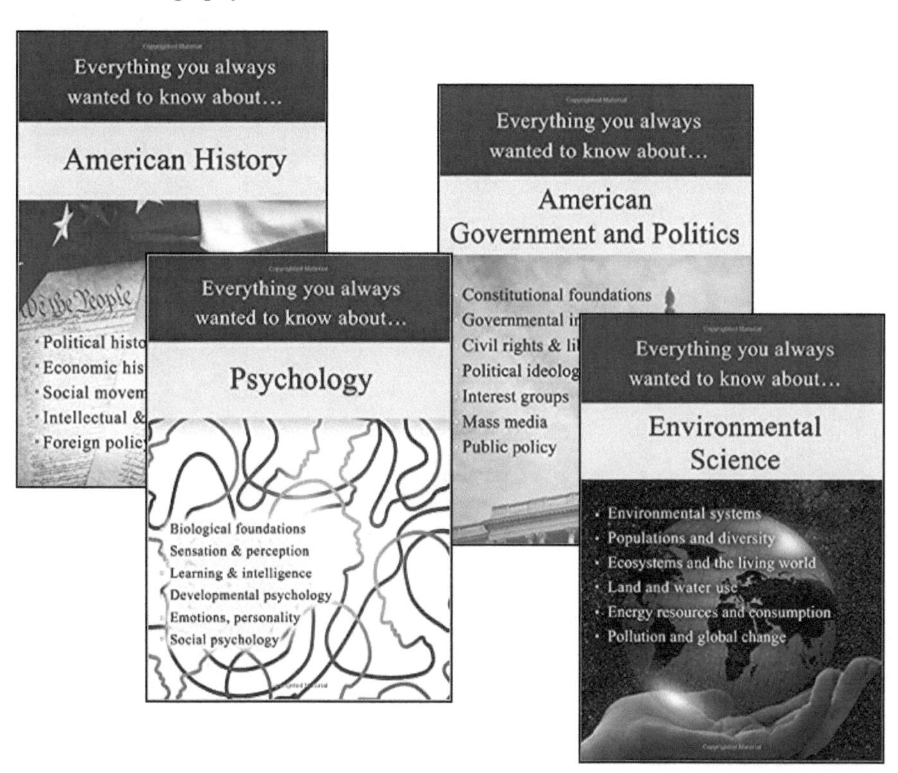

Page intentionally left blank

Table of Contents

REVIEW: Acids & Bases *(continued)*

REVIEW

Acids & Bases

- Acid–Base Nomenclature
- Acidic and Basic Solutions
- Acid–Base Equilibria
- Common Ions and Salts
- Acid–Base Ionization
- Periodic Trends of Acids and Bases
- Percent Dissociation
- Salts of Acid–Base Reactions
- pH calculations for Acid–Base Solutions
- Buffered Solutions
- Ionic Equations and Neutralization
- Titration
- Indicators
- Titration Curves

Acid–Base Nomenclature

Characteristics of acids and bases

Before the twentieth century, acids, bases, and salts were characterized by properties such as taste and the ability to change litmus's color (a water-soluble mixture of organic dyes).

Acids taste *sour* (e.g., lemon juice), bases taste *bitter* (e.g., mustard), and salts, as their name suggests, taste *salty* (e.g., sodium chloride as table salt).

Acids cause blue litmus paper to turn *red*, bases turn red litmus paper *blue*, while neutral compounds do not affect litmus paper's color.

Bases are recognized by their slippery feel (e.g., soap).

The position on the pH scale categorically classifies acids and bases.

Some complex metabolic processes in the human body are controlled by physiological pH (≈ 7.4); a small change may lead to severe illness and death.

The chemistry of acids and bases has a significant role in nature and industry processes. For example, soil acidity is essential for plant growth.

Acids and bases are particularly essential in manufacturing industries. Sulfuric acid (H_2SO_4) is the most widely produced chemical, and it is needed to produce fertilizers, polymers, steel, and many other materials.

The extensive use of sulfuric acid has led to environmental problems, such as the phenomenon of acid rain.

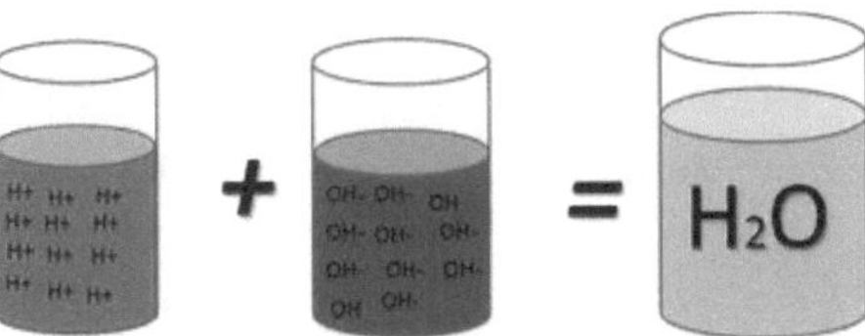

Acids containing protons (H^+) *Bases containing hydroxides (^-OH)* *Acids and bases combine to produce water (H_2O)*

Binary acids and oxoacids

There are *two types of acids*:

binary acids (acids that do not contain oxygen atoms)

oxoacids (acids containing oxygen atoms).

Acids have slightly different names from the ionic naming rules discussed, while bases follow the ionic naming rules discussed.

Naming binary acids

Binary acids start with *hydro-*, then the first syllable of the anion's name, and end with *~ic*:

HF : *hydro*fluor*ic* acid

HCl : *hydro*chlor*ic* acid

HI : *hydro*iod*ic* acid

HBr : *hydro*brom*ic* acid

H_2S : *hydro*sulfur*ic* acid

HCN : *hydro*cyan*ic* acid

Oxoacid's names derive from the name of the oxyanion of the acid.

Naming associated anions

For anions ending with *~ate*, the acid's name starts with the first syllable of the anion name and ends with *~ic*.

For anions ending with *~ite*, the name of acid starts with the first syllable of the anion's name and ends with *-ous*.

Anion	Name of Anion	Acid	Name of Acid
NO_3^-	nitrate ion	HNO_3	nitric acid
NO_2^-	nitrite ion	HNO_2	nitrous acid
SO_4^{2-}	sulfate ion	H_2SO_4	sulfuric acid
SO_3^{2-}	sulfite ion	H_2SO_3	sulfurous acid
PO_4^{3-}	phosphate ion	H_3PO_4	phosphoric acid
$C_2H_3O_2^-$	acetate ion	$HC_2H_3O_2$	acetic acid
ClO^-	hypochlorite	HClO	hypochlorous acid
ClO_2^-	chlorite	$HClO_2$	chlorous acid
ClO_3^-	chlorate	$HClO_3$	chloric acid
ClO_4^-	perchlorate	$HClO_4$	perchloric acid

Arrhenius acids and bases

In 1884, Swedish chemist Svante Arrhenius defined acids and bases:

> *Acids* dissociate in water to produce *hydrogen ions* (H^+)
>
> *Bases* dissociate to in water produce *hydroxide ions* (^-OH)

For example, HCl is an *acid*:

$$HCl\ (aq) \rightarrow H^+\ (aq) + Cl^-\ (aq)$$

For example, NaOH is a *base*:

$$NaOH\ (aq) \rightarrow Na^+\ (aq) + {}^-OH\ (aq)$$

Arrhenius acids increase the hydronium ion concentration [H_3O^+] in an aqueous solution.

Arrhenius bases increase the hydroxide ion concentration [^-OH] in an aqueous solution.

Acids, according to the Arrhenius concept:

1. $HCl\ (aq) + H_2O \rightarrow H_3O^+\ (aq) + Cl^-\ (aq)$

2. $HNO_3\ (aq) + H_2O \rightarrow H_3O^+\ (aq) + NO_3^-\ (aq)$

3. $CH_3COOH\ (aq) + H_2O \rightleftarrows H_3O^+\ (aq) + CH_3COO^-\ (aq)$

Bases, according to the Arrhenius concept:

1. $NaOH\ (aq) \rightarrow Na^+\ (aq) + OH^-\ (aq)$

2. $Ba(OH)_2\ (aq) \rightarrow Ba^{2+}\ (aq) + 2\ OH^-\ (aq)$

3. $NH_3\ (aq) + H_2O \rightleftarrows NH_4^+\ (aq) + OH^-\ (aq)$

Lewis acids and bases

A *Lewis base* provides *a lone pair of electrons* to another reactant, forming a coordinate covalent bond.

A *Lewis acid* has an empty orbital to accept *a pair of electrons* from another reactant to form a coordinate covalent bond.

In 1923, Gilbert Newton Lewis (1875-1946) proposed a definition that Lewis acids (e.g., H^+) accept a pair of nonbonding electrons.

Lewis acids are electron-pair *acceptors.*

Bases are electron-pair *donors.*

H+ is the acid while water and ammonia are the Lewis bases in the following reactions:

H^+	+	H_2O	$\rightarrow$	H_3O^+
Lewis acid		Lewis base		

H^+	+	NH_3	$\rightarrow$	NH_4^+
Lewis acid		Lewis base		

In reactions forming new covalent bonds, the species with an incomplete octet (i.e., an electron-deficient molecule) may act as Lewis acids; a lone pair of electrons may act as Lewis bases.

In the following reactions,

BF_3, $AlCl_3$, and $FeBr_3$ are Lewis acids

while

NH_3, Cl^- and Br^- are Lewis bases

$BF_3 + NH_3 \rightarrow F_3B{:}NH_3$ $\qquad AlCl_3 + Cl^- \rightarrow AlCl_4^-$ $\qquad FeBr_3 + Br^- \rightarrow FeBr_4^-$

In the formation of complex ions, the positive ions act as Lewis acids, and the ligands (anions or small molecules) act as Lewis bases:

$Cu^{2+}(aq)$	+	$4\ NH_3\ (aq)$	$\rightleftarrows$	$Cu(NH_3)_4^{2+}\ (aq)$
Lewis acid		Lewis base		

$Al^{3+}(aq)$	+	$6\ H_2O$	$\rightleftarrows$	$[Al(H_2O)_6]^{3+}\ (aq)$
Lewis acid		Lewis base		

The *ionizable hydrogen* in oxoacids is bonded to the oxygen.

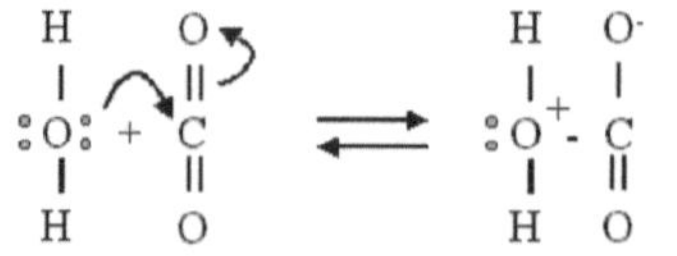

Lewis base *Lewis acid*

Brønsted–Lowry acids and bases

The *Brønsted-Lowry theory* was proposed, in 1923, independently by Danish chemist Johannes Nicolaus Brønsted (1879-1947) and British chemist Martin Lowry (1874-1936).

The Brønsted-Lowry theory states that:

an acid is a substance that acts as a proton donor, and

a base is a substance that acts as a proton acceptor.

Brønsted–Lowry conjugate acids and conjugate bases

From the exchange of protons,

acids form its *conjugate base*

bases form its *conjugate acid*

Brønsted-Lowry acid-base reactions can be represented as follows:

HA	+	B	$\rightleftarrows$	BH^+	+	A^-
acid		base		conjugate acid		conjugate base

Brønsted-Lowry acids, bases, conjugate acids, and conjugate bases

HCl	+	H_2O	$\rightarrow$	H_3O^+ (*aq*)	+	Cl^- (*aq*)
acid		base		conjugate acid		conjugate base

$HC_2H_3O_2$	+	H_2O	$\rightleftarrows$	H_3O^+ (*aq*)	+	$C_2H_3O_2^-$ (*aq*)
acid		base		conjugate acid		conjugate base

NH_3	+	H_2O	$\rightleftarrows$	NH_4^+ (*aq*)	+	OH^- (*aq*)
base		acid		conjugate acid		conjugate base

The transfer of protons in a Brønsted-Lowery acid-base reaction

HF (*aq*)	+	NH_3	$\rightleftarrows$	F^-	+	$^+NH_4$ (*aq*)
Acid donates H^+ to NH_3		Base accepts H^+ from HF		Conjugate base accepts H^+ / $^+NH_4$		Conjugate acid donates H^+ to F^-

Amphoteric substances

Water is an *amphoteric* substance because it can *act as an acid or a base*. It acts as a base by accepting a hydrogen ion; another water molecule acts as an acid.

Therefore, the ions H_3O^+ (*aq*) + OH^- (*aq*) form in pure water.

As ions form, they react to produce water again with the following equilibrium:

H_2O +	H_2O ⇄	H_3O^+ (*aq*) +	OH^- (*aq*)
acid	base	conjugate acid	conjugate base

Autoionization of water

In the autoionization of water, a proton is transferred from one water molecule to another, producing a hydronium ion (H_3O^+) and a hydroxide ion (OH^-).

Ions formed in the auto-ionization of water are:

$$H_2O\ (l) + H_2O\ (l) \rightleftarrows H_3O^+\ (aq) + OH^-\ (aq)$$

This process of exchange is the *autoionization of water*.

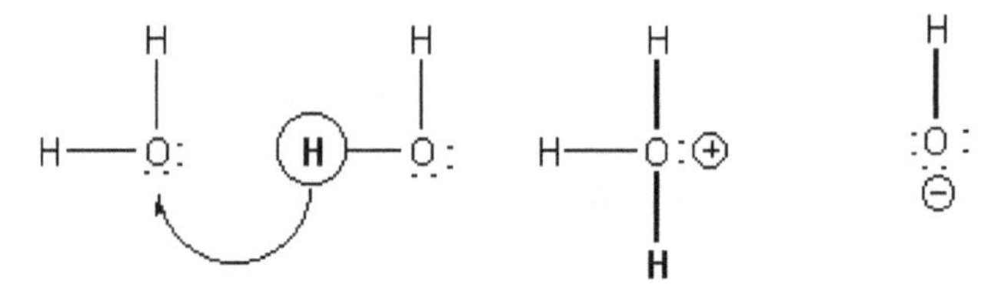

H⁺ acceptor base　*H⁺ donor acid*　*Conjugate acid*　*Conjugate base*

Equilibrium constant K_w for autoionization

The equilibrium constant *K*w:

$$K_w = [H_3O^+]\cdot[^-OH]$$

where *K*w is the autoionization constant for water

At standard temperature and pressure (25 °C, 1 atm), the equilibrium constant K_w (*ion-product constant*) for water:

$$K_w = [H_3O^+]\cdot[^-OH]$$

$$K_w = 1.0 \times 10^{-14}$$

For example,

If $[H_3O^+]$ increases ($> 1.0 \times 10^{-7}$ *M*)

If $[^-OH]$ decreases ($< 1.0 \times 10^{-7}$ M)

If $[H_3O^+] = [^-OH] = 1.0 \times 10^{-7}$ M, neutral solution (e.g., pure water)

If $[H_3O^+] > 1.0 \times 10^{-7}$ M, $[^-OH] < 1.0 \times 10^{-7}$ M, solution is acid ($[H^+] > [^-OH]$)

If $[H_3O^+] < 1.0 \times 10^{-7}$ M, $[^-OH] > 1.0 \times 10^{-7}$ M, solution is basic ($[H^+] < [^-OH]$)

Notes for active learning

Acidic and Basic Solutions

pH and pOH

The pH scale measures acidity or basicity, especially when the hydrogen ion (H^+) concentration is extremely low.

$$pH = -\log[H^+]$$

$$pOH = -\log[^-OH]$$

Neutral solutions

$$[H^+] = 1.0 \times 10^{-7}\ M$$

$$pH = -\log(1.0 \times 10^{-7})$$

$$pH = 7.00$$

Neutral solutions contain

$$[^-OH] = 1.0 \times 10^{-7}\ M$$

$$pOH = -\log(1.0 \times 10^{-7})$$

$$pOH = 7.00$$

Acidic solutions

$$[H^+] > 1.0 \times 10^{-7}\ M$$

$$pH < 7.00$$

Basic solutions

$$[H^+] < 1.0 \times 10^{-7}\ M$$

$$pH > 7.00$$

pH standards

$pH = 7 \rightarrow$ a neutral solution

$pH < 7 \rightarrow$ an acidic solution

$pH > 7 \rightarrow$ a basic solution

Note

$$K_w = [H^+]\cdot[^-OH]$$

$$K_w = 1.0 \times 10^{-14}$$

$$pK_w = -\log(K_w)$$

$$pK_w = -\log(1.0 \times 10^{-14})$$

$$pK_w = 14.00$$

but

$$pK_w = pH + pOH = 14.00$$

$$pOH = 14.00 - pH$$

$$pH = 14.00 - pOH$$

if

$$pH = 7$$

$$pOH = 7$$

if

$$pH < 7$$

$$pOH > 7$$

if

$$pH > 7$$

$$pOH < 7$$

thus

$$pH < 7$$

$$[H^+] > [^-OH]$$

and

$$pH > 7$$

$$[^-OH] > [H^+]$$

For example. if

$[H^+] = 1.0 \times 10^{-4}$ M

$pH = -\log(1.0 \times 10^{-4})$

pH = 4.00

when

$[^-OH] = 1.0 \times 10^{-4}$ M

$[H^+] = 1.0 \times 10^{-14} / 1.0 \times 10^{-4}$ M

$[H^+] = 1.0 \times 10^{-10}$ M

$pH = -\log(1.0 \times 10^{-10})$

pH = 10.00

Alternatively,

$[^-OH] = 1.0 \times 10^{-4}$ M

$pOH = -\log[^-OH]$

$pOH = -\log(1.0 \times 10^{-4}\text{ M})$

pOH = 4.00

pH = 14.00 – 4.00

pH = 10.00

Reference values for $[H^+]$ and pH

$[H^+]$, M	pH	$[H^+]$, M	pH
1.0×10^{-1}	1.00	1.0×10^{-8}	8.00
1.0×10^{-2}	2.00	1.0×10^{-9}	9.00
1.0×10^{-3}	3.00	1.0×10^{-10}	10.00
1.0×10^{-4}	4.00	1.0×10^{-11}	11.00
1.0×10^{-5}	5.00	1.0×10^{-12}	12.00
1.0×10^{-6}	6.00	1.0×10^{-13}	13.00
1.0×10^{-7}	7.00	1.0×10^{-14}	14.00

pH of an acidic solution

To calculate the pH of an acidic solution, use the expression:

$$pH = -\log[H_3O^+]$$

For example, if

$$[H_3O^+] = 1.0 \times 10^{-2}\ M$$

$$pH = -\log(1.0 \times 10^{-2})$$

$$pH = -\ (-2.00)$$

$$pH = 2.00\ (\rightarrow \text{acidic})$$

pH of a basic solution

To calculate the pH of a basic solution, use the expression:

$$pOH = -\log[^-OH]$$

If a solution has

$$[OH^-] = 1.0 \times 10^{-2}\ M$$

$$pOH = -\log(1.0 \times 10^{-2})$$

$$pOH = -\ (2.00)$$

$$pOH = 2.00\ (\rightarrow \text{basic})$$

Since at 25 °C

$$K_w = [H_3O^+]\cdot[OH^-]$$

$$K_w = 1.0 \times 10^{-14}$$

$$pK_w = -\log(K_w)$$

$$pK_w = -\log[H_3O^+] + (-\log[^-OH])$$

$$pK_w = -\log(1.0 \times 10^{-14})$$

$$pK_w = -\ (-14.00)$$

continued...

$pK_w = 14.00$

$pK_w = pH + pOH$

$pK_w = 14.00$

$pOH = 14.00 - pH$

Thus, in aqueous solutions

$pH = 2$

$pOH = 12$

and

$pOH = 2$

$pH = 12$

Notes for active learning

Acid–Base Equilibria

Equilibrium constants K_a and K_b

Equilibrium constants K_a and K_b measure the extent to which an acid or base dissociates (dissociation constants).

The strength of an acid is defined by its dissociation (ionization) in an *aqueous* (*aq*) solution.

$$HX\ (aq) + H_2O \rightleftarrows H_3O^+\ (aq) + X^-\ (aq)$$

The equilibrium constant, K_a, for the acid ionization:

$$K_a = [H_3O^+]\cdot[X^-]\ /\ [HX]$$

For K_a, the products (conjugate acid H_3O^+ and conjugate base X^-) of the dissociation are the numerator, while the parent acid (HX) is the denominator.

The strength of a base is defined by its dissociation (ionization) in an *aqueous* (*aq*) solution.

$$X^-\ (aq) + H_2O \rightleftarrows OH^-\ (aq) + HX\ (aq)$$

$$K_b = [HX^+]\cdot[^-OH]\ /\ [X]$$

For K_b, the products (conjugate acid HX^+ and conjugate base OH^-) of the dissociation are the numerator, while the parent base (X) is the denominator.

Acid dissociation

The value of K_a measures the extent of *acid dissociation*, hence the acid's relative strength.

For strong acids (e.g., $HClO_4$, HCl, H_2SO_4, HNO_3), K_a is exceptionally large (not in the reported K_a values).

For weak acids, $K_a << 10^{-1}$.

Strong bases have larger K_b values.

Weak bases have exceedingly small K_b values.

If the K_a value for a conjugate acid-base pair is known, the K_b can be calculated (and vice versa) using the following relationship:

$$K_aK_b = K_w$$

Sodium acetate dissolves in water dissociating into sodium (Na^+) and acetate ions ($C_2H_3O_2^-$):

$$NaC_2H_3O_2\ (aq) \rightarrow Na^+\ (aq) + C_2H_3O_2^-\ (aq)$$

If the K_a is 1.8×10^{-5} (provided in a reference table), what is the K_b?

The acetate ion reacts with water with the following equilibrium:

$$C_2H_3O_2^- (aq) + H_2O \rightleftarrows HC_2H_3O_2 (aq) + OH^- (aq)$$

$$K_b = [HC_2H_3O_2]\cdot[OH^-] / [C_2H_3O_2^-]$$

For the dissociation of acetic acid:

$$HC_2H_3O_2 (aq) + H_2O \rightleftarrows H_3O^+ (aq) + C_2H_3O_2^- (aq)$$

$$K_a = [H_3O^+]\cdot[C_2H_3O_2^-] / [CH_3COOH]$$

$$K_a \times K_b = [H_3O^+]\cdot[C_2H_3O_2^-] / [CH_3COOH] \times [HC_2H_3O_2]\cdot[OH^-] / [C_2H_3O_2^-]$$

$$K_w = [H_3O^+]\cdot[OH^-]$$

$$K_w = 1.0 \times 10^{-14}$$

Thus, for acetate ion $C_2H_3O_2^-$ in solution

$$K_b = K_w / K_a \text{ (for } HC_2H_3O_2)$$

$$K_b = (1.0 \times 10^{-14}) / (1.8 \times 10^{-5})$$

$$K_b = 5.6 \times 10^{-10} (> K_w)$$

An aqueous solution of 0.10 M $NaC_2H_3O_2$ has $[OH^-] \sim 7.5 \times 10^{-6}$ M and pH ~ 8.9.

pK_a and pK_b measure acidity and basicity

The operator "p" means "take the negative logarithm of."

Therefore

$$pK_a = -\log K_a$$

As K_a gets larger, pK_a gets smaller.

The *smaller* the pK_a, the *stronger* the acid (e.g., pK_a for H_2SO_4 = 1.92).

$$pK_b = -\log K_b$$

As K_b gets larger, pK_b gets smaller.

The *smaller* the value of pK_b, the *stronger* the base.

Conjugate acids and bases

A *conjugate base* is the species remaining from the acid after it *loses* a proton.

A *conjugate acid* is the species remaining from the base after *gaining* a proton.

The *conjugate acid-base pairs* ($acid_1$–conjugate $base_1$ and $acid_2$–conjugate $base_2$) are related substances by the loss or gain of a single proton (H^+).

H_2O and H_3O^+, and H_2O and OH^- are conjugate acid-base pairs, but H_3O^+ and OH^- are not.

Strong acids have *weak conjugate bases.*

Weak acids have *strong conjugate bases.*

The *weaker the acid*, the stronger its conjugate base.

Strong bases have *weak conjugate acids.*

Weak bases have *strong conjugate acids.*

The *weaker the base*, the stronger the conjugate acid it produces.

HCl is a strong acid, and Cl^- is a weak base.

HF is a weak acid, and F^- is a strong conjugate.

A Brønsted-Lowry acid-base reaction involves a competition between two bases for a proton, in which the stronger base is the most protonated at equilibrium.

$$HCl + H_2O \rightarrow H_3O^+ (aq) + Cl^- (aq)$$

H_2O is a stronger base than the Cl^-.

At equilibrium, the HCl solution contains mainly H_3O^+ and Cl^- ions.

$$HC_2H_3O_2 (aq) + H_2O \rightleftarrows H_3O^+ (aq) + C_2H_3O_2^- (aq)$$

$C_2H_3O_2^-$ is the stronger base.

At equilibrium, acetic acid contains mainly $HC_2H_3O_2$ and few H_3O^+ and $C_2H_3O_2^-$ ions.

$$NH_3 (aq) + H_2O \rightleftarrows NH_4^+ (aq) + {}^-OH (aq)$$

H_2O is an acid. Competition for protons occurs between NH_3 and ^-OH, where ^-OH is the stronger base, and the equilibrium favors the reactants. An aqueous ammonia solution contains *mostly NH_3 molecules* and smaller amounts of NH_4OH, NH_4^+, and ^-OH.

According to Brønsted-Lowry, the net acid-base reactions favor *strong acid-strong base* combinations to *weak acid-weak base* combinations.

Acid-base reaction equilibrium

The following acid-base reactions proceed in the forward direction:

$HCl\ (aq) + NH_3\ (aq) \rightarrow NH_4^+\ (aq) + Cl^-\ (aq)$

$HSO_4^-\ (aq) + CN^-\ (aq) \rightarrow HCN\ (aq) + SO_4^{2-}\ (aq)$

HSO_4^- ($pK_a = 6.91$) is a stronger acid than HCN ($pK_a = 9.21$)

Many acid-base reactions reach a state of equilibrium.

For the following acid-base reactions, the equilibrium may favor the products or the reactants, depending on the relative strength of the acid:

$H_2PO_4^-\ (aq) + C_2H_3O_2^-\ (aq) \rightleftarrows HC_2H_3O_2\ (aq) + HPO_4^{2-}\ (aq)$

Equilibrium shifts to the *left* (reactants):

$HC_2H_3O_2$ ($pK_a = 4.75$) is the stronger acid

HPO_4^{2-} ($pK_a = 7.21$) is the stronger base

$HNO_2\ (aq) + C_2H_3O_2^-\ (aq) \leftrightarrows HC_2H_3O_2\ (aq) + NO_2^-\ (aq)$

Equilibrium shifts to the *right* (products).

HNO_2 ($pK_a = 3.39$) is the stronger acid

$C_2H_3O_2^-$ ($pK_a = 4.75$) is the stronger base

Acid–Base Ionization

Strength of acids and bases

A *strong acid* ionizes entirely in aqueous solutions.

Strong acids include $HClO_4$, HCl, H_2SO_4, HNO_3, HBr, and HI.

For these strong acids, the equilibrium lies *far to the right.*

Weak acids partially ionize with ionization equilibriums *far to the left.*

Weak acids include:

$HC_2H_3O_2$, HNO_2, H_2SO_3, H_3PO_4, and $HClO$.

For example, consider the reversible process when an acid dissolves in water:

$$HA\ (aq) + H_2O \leftrightarrow H_3O^+\ (aq) + A^-\ (aq)$$

In the forward reaction above, the acid HA donates a proton to the water molecule to form a hydronium ion (H_3O^+) and conjugate base (A^-).

Water acts as a Brønsted-Lowry base, and the acid strength is measured by its degree of ionization (or dissociation) in water.

If an acid ionizes completely, it is a strong acid.

A *strong acid* has a *weak conjugate base* (i.e., the conjugate base loses its proton to water readily). Cl^-, Br^-, I^-, ClO_4^-, HSO_4^-, and NO_3^- are weak conjugate bases.

A strong acid is less able to compete with water for a proton.

Ionization constant K_a

The ionization constant K_a for a *strong acid is exceptionally large*, and the equilibrium shifts far to the right (i.e., dissociated proton and stable anion).

A weak acid does not readily give up its proton to water, and it has a strong conjugate base.

Strong conjugate bases include:

$C_2H_3O_2^-$, F^-, CN^-, NO_2^-, HSO_3^-, SO_3^{2-}, $H_2PO_4^-$, HPO_4^{2-}, PO_4^{3-}.

The weaker the acid, the stronger its conjugate base.

The ionization equilibrium for weak acids shifts far to the left.

An aqueous solution of a strong acid has hydronium ions (H_3O^+) and the acid's conjugate base.

For example, there are H_3O^+ and Cl^- ions in an aqueous HCl solution, but virtually no HCl molecules.

An aqueous solution of a weak acid, acetic acid, contains mainly undissociated molecules, $HC_2H_3O_2$, with about ~1% H_3O^+ and $C_2H_3O^-$ ions.

Bond strength and polarity for acids

Factors determine the strength of acids, such as the *strength* and *polarity* of the X–H bond in the molecule and the hydration energy of the ionic species in an aqueous solution.

For inorganic binary acids, H–X bond strength decreases down the group:

HF, HCl, HBr, HI),

The weaker the bond, the easier it ionizes in an aqueous solution.

The stronger bond is the acid; strength increases down the group:

$HF < HCl < HBr < HI$

Relative ionization strength

Among the hydrohalic acids, HF is the weak acid; the others are strong acids. The relative strength of HCl, HBr, and HI cannot be differentiated in an aqueous solution because each of them dissociates almost completely.

Less polar solvents are used to determine their relative strength.

Strong bases, like strong acids (e.g., NaOH, KOH, and $Ba(OH)_2$), ionize completely when dissolved in water. A strong base tends to accept a proton.

Weak bases do not ionize completely when dissolved in water and show a minimal tendency to accept a proton.

HCl, HBr, and HI ionize partially in acetone or methanol, which have a weaker ionizing strength than water.

The ionization of HCl in acetone:

$(CH_3)_2CO$ (*l*) (*acetone*) + HCl $\rightleftarrows$ $(CH_3)_2COH^+ + Cl^-$

The degree of ionization in acetone increases as:

$HCl < HBr < HI$

Oxoacids produce H+

Oxoacids contain one or more ~OH groups covalently bonded to a central atom, a metal, or a nonmetal.

The ~OH group ionizes entirely or partially in an aqueous solution, producing hydrogen ions.

Examples of oxoacids: H_2CO_3, HNO_3, H_3PO_4, H_2SO_4, $HClO_4$, $HC_2H_3O_2$

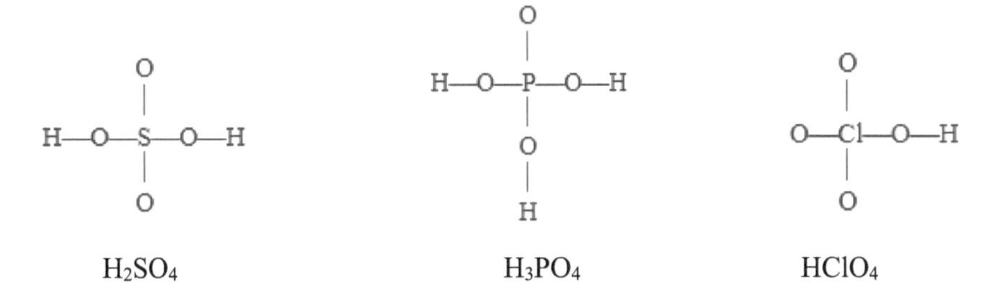

Notes for active learning

Periodic Trends of Acids and Bases

Periodic trends for acidity

The acidity of hydrogen halides increases down a group; trend of relative acidity for hydrides of Group 15:

$H_2O < H_2S < H_2Se < H_2Te$

For the same period hydrides, relative acidity increases from left to right:

$CH_4 < NH_3 < H_2O < HF$

$PH_3 < H_2S << HCl$

Water is a stronger acid than ammonia, and in an acid-base reaction, H_2O acts as a Brønsted-Lowry acid, which donates a proton to NH_3:

$H_2O + NH_3\ (aq) \rightleftarrows NH_4^+\ (aq) + OH^-\ (aq)$

In reaction with HF, water acts as a Brønsted-Lowry base, which accepts a proton:

$HF\ (aq) + H_2O \rightleftarrows H_3O^+\ (aq) + F^-\ (aq)$

Electronegativity for acid strength

Relative acid strength depends on the central atoms' *electronegativity*.

The more electronegative, the more polarized the O–H bond, and the *more readily it ionizes* in an aqueous solution to release the H^+ ion.

For example, N, S, and Cl are more electronegative than P; HNO_3, H_2SO_4, and $HClO_4$ are stronger acids, whereas H_3PO_4 is weak.

The order of electronegativity for relative acid strength:

$Cl \approx N > S > P$

Acid strength:

$HClO_4 > HNO_3 > H_2SO_4 > H_3PO_4$

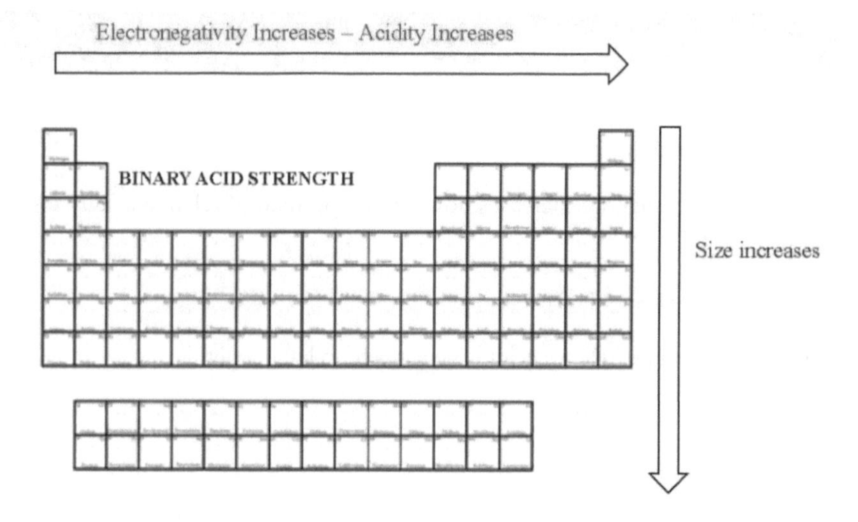

Acidity increases with increases in electronegativity and atomic size

Induction and acid strength

For oxoacids that have the central atoms with elements of the same group in the periodic table, relative strength decreases from top to bottom (as the electronegativity of the central atom decreases):

$HOCl > HOBr > HOI$

$HClO_2 > HBrO_2 > HIO_2$

$HClO_4 > HBrO_4 > HIO_4$

For oxoacids containing identical central atoms, acidity increases as oxygen atoms bond.

Acidity increases as follows:

$HOCl < HClO_2 < HClO_3 < HClO_4$

$H_2SO_3 < H_2SO_4$

$HNO_2 < HNO_3$

When oxygen atoms bond to the central atom, the O–H bond in the molecule becomes highly polarized (due to the inductive electronegative effect).

The O–H bond ionizes readily to release an H^+ ion.

Acetic acid (CH_3COOH) is an organic acid containing the carboxyl (~COOH) group.

In an aqueous solution, ionization of an acetic acid involves breaking the carboxyl group's O–H bond, but not the C–H bonds in the methyl group (CH_3).

However, if one or more of the hydrogen atoms in the methyl group is substituted with a more electronegative atom, the inductive effect causes the electron cloud to be drawn away from the carbonyl group. The O–H bond becomes more polarized and ionizes readily, increasing the acidity.

The following K_a values illustrate the effect on acetic acid's acidity and its derivatives when methyl hydrogens are substituted with electronegative atoms.

The stronger the acid, the higher the K_a value:

CH_3COOH (*aq*) (*acetic acid*) + $H_2O \rightleftarrows CH_3COO^-$ (*aq*) + H_3O^+ (*aq*)

$K_a = \mathbf{1.8 \times 10^{-5}}$

$ClCH_2COOH$ (*aq*) (*chloroacetic acid*) + $H_2O \rightleftarrows ClCH_2COO^-$ (*aq*) + H_3O^+ (*aq*)

$K_a = \mathbf{1.4 \times 10^{-3}}$

FCH_2COOH (*aq*) (fluoroacetic acid) + $H_2O \rightleftarrows FCH_2COO^-$ (*aq*) + H_3O^+ (*aq*)

$K_a = \mathbf{2.6 \times 10^{-3}}$

CCl_3COOH (*aq*) (trichloroacetic acid) + $H_2O \rightleftarrows CCl_3COO^-$ (*aq*) + H_3O^+ (*aq*)

$K_a = \mathbf{3.0 \times 10^{-1}}$

Monoprotic and polyprotic acids

The *monoprotic acids* include HCl, HF, HOCl, HNO_2, and $HC_2H_3O_2$ because each molecule contains a single ionizable hydrogen ion.

The *polyprotic acids* contain more than one ionizable hydrogen, and examples include:

H_2SO_4, H_2SO_3, $H_2C_2O_4$ and H_3PO_4

The hydrogen ionizes in stages with ionization constants, such as for H_3PO_4.

H_3PO_4 (*aq*) $\rightleftarrows H^+$ (*aq*) + $H_2PO_4^-$ (*aq*); $K_{a1} = 7.5 \times 10^{-3}$

$H_2PO_4^-$ (*aq*) $\rightleftarrows H^+$ (*aq*) + HPO_4^{2-} (*aq*); $K_{a2} = 6.2 \times 10^{-8}$

HPO_4^{2-} (*aq*) $\rightleftarrows H^+$ (*aq*) + PO_4^{3-} (*aq*); $K_{a3} = 4.8 \times 10^{-13}$

From above, acid strength decreases in order:

$H_3PO_4 >> H_2PO_4^- >> HPO_4^{2-}$

Ionization of polyprotic acids

Sulfuric acid (H_2SO_4) is a strong acid, but only the first hydrogen ionizes completely:

$$H_2SO_4\,(aq) \rightarrow H^+\,(aq) + HSO_4^-\,(aq)$$

$$K_{a1} = \text{exceptionally large}$$

The second H does not often dissociate, and HSO_4^- is a weak acid:

$$HSO_4^-\,(aq) \rightleftarrows H^+\,(aq) + SO_4^{2-}\,(aq)$$

$$K_{a2} = 1.2 \times 10^{-2}$$

Hydroxide bases

Weak bases include NH_3 (or NH_4OH), NH_2OH, $Mg(OH)_2$, and hydroxides and oxides as slightly soluble in water.

Hydroxides of Group I metals (LiOH, NaOH, KOH, ROH, and CsOH) are strong bases, but only NaOH and KOH are commercially important and commonly used laboratory bases.

Hydroxide bases are *soluble* in water, and they dissociate entirely in an aqueous solution, producing a high concentration of hydroxide ions.

A moderately dilute solution of NaOH contains [^-OH], [Na+], and [NaOH].

$Ba(OH)_2$ is a relatively strong base among the alkaline earth metal hydroxides.

The other metal hydroxides are sparingly soluble in water, limiting their basicity.

A saturated solution of these hydroxides contains a low concentration of ^-OH.

These hydroxides react with strong acids:

$$Na_2O\,(s) + H_2O \rightarrow 2\,NaOH\,(aq)$$

$$BaO\,(s) + H_2O \rightarrow Ba(OH)_2\,(aq)$$

$$MgO\,(s) + 2\,HCl\,(aq) \rightarrow MgCl_2\,(aq) + H_2O$$

Hydroxides of some metals (e.g., $Al(OH)_3$, $Cr(OH)_3$, $Zn(OH)_2$, $Sn(OH)_2$ and $Pb(OH)_2$) exhibit amphoteric properties (i.e., act as an acid or a base).

For example:

$$Al(OH)_3\,(s) + OH^-\,(aq) \rightleftarrows Al(OH)_4^-\,(aq)$$

$$Al(OH)_3\,(s) + 3\,H_3O^+\,(aq) \rightleftarrows [Al(H_2O)_6]^{3+}\,(aq)$$

Hydride bases

Hydrides of reactive metals (e.g., NaH, MgH_2, CaH_2) form strongly basic solutions if dissolved in water.

The hydride ion reacts with water to produce hydroxide ions and hydrogen gas:

$$H^- (aq) + H_2O \rightarrow H_2 (g) + OH^- (aq)$$

The oxide ion O^{2-} has an extraordinarily strong affinity for protons (H^+), reacting with H_2O to form hydroxide ions.

$$O^{2-} (aq) + H_2O \rightarrow 2\ OH^- (aq)$$

Oxides of nonmetals are acidic, forming acidic solutions when dissolved in water.

$$CO_2 (g) + H_2O \rightleftarrows H_2CO_3 (aq) \rightleftarrows H^+ (aq) + HCO_3^- (aq)$$

$$SO_2 (g) + H_2O \rightleftarrows H_2SO_3 (aq) \rightleftarrows H^+ (aq) + HSO_3^- (aq)$$

Ammonia is the only weak base that is of commercial importance.

Ammonia does not contain hydroxide ions, but it reacts with water and ionizes as follows:

$$NH_3 (aq) + H_2O \rightleftarrows NH_4^+ (aq) + OH^- (aq)$$

The base dissociation constant K_b is given by the expression:

$$K_b = [NH_4^+]\cdot[^-OH] / [NH_3]$$

$$K_b = 1.8 \times 10^{-5}$$

Notes for active learning

Percent Dissociation

Degree of ionization

The *degree of ionization* (or *percent dissociation*) of a weak acid is:

Percent dissociation = [acid ionized] / [initial acid] × 100%

For strong acids, the percent dissociation at equilibrium is ~ 100%.

The percent dissociation for weak acids depends on the acid's K_a and initial concentration.

For example, the percent dissociation of acetic acid (Use $HC_2H_3O_2$ $K_a = 1.8 \times 10^{-5}$) at 0.10 M concentration is:

$(1.3 \times 10^{-3}$ M / 0.10 M) × 100% = 1.3 %

The *stronger* the acid, the larger is the K_a, and the *greater* is the percent ionization.

The *percent ionization of a weak acid* depends on the K_a and the extent of dilution.

The *more dilute* an acid solution, the *higher* is the percentage of ionization.

For example, consider a 0.010 M acetic acid solution and its ionization products.

Use an ICE table (**I**nitial, **C**hange, **E**quilibrium) to simplify calculations in reversible equilibrium reactions.

Once the equilibrium row is completed (by summing the initial and change rows), its contents can be substituted into the equilibrium constant expression to solve for K_a.

$CH_3COOH\ (aq) \rightleftarrows H^+\ (aq) + CH_3COO^-\ (aq)$

Initial [ ], M:	0.010	0.00	0.00
Change, Δ[ ], M:	$-x$	$+x$	$+x$
Equilibrium [ ], M :	$(0.010 - x)$	x	x

Acid ionization constant K_a

The acid ionization constant, K_a, is given by:

$K_a = [H_3O^+] \cdot [C_2H_3O_2^-] / [CH_3COOH]$

$K_a = x^2 / (0.010 - x)$

$K_a = 1.8 \times 10^{-5}$

Since $K_a << 0.010$, approximate that $x << 0.010$, and $(0.010 - x) \sim 0.010$

Then

$$K_a = x^2 / (0.010 - x)$$

$$K_a \approx x^2 / 0.010$$

$$K_a = 1.8 \times 10^{-5}$$

$$x^2 = 1.8 \times 10^{-7}$$

and

$$x = \sqrt{(1.8 \times 10^{-7})}$$

$$x = 4.2 \times 10^{-4}$$

$$x = [H_3O^+]$$

$$x = 4.2 \times 10^{-4}\ M$$

The *degree of ionization* of acetic acid at this concentration is

$$(4.2 \times 10^{-4}\ M / 0.010\ M) \times 100\% = 4.2\%$$

The degree of ionization of the acid increases as the solution is diluted. For 0.1 M acetic acid, the degree of ionization is 0.42%, 10-fold lower than in 0.010 M acid solution.

The percent dissociation of weak bases depends on the base solution's K_b value and dilution.

Larger K_b and greater dilution result in a higher percent dissociation.

Hydrolysis of salts

The presence of salt affects the dissociation of acids and bases due to salts' *hydrolysis*.

For example, CH_3COOH dissociates less in a solution containing CH_3COONa salt, while NH_4OH dissociates less in a solution containing NH_4Cl salt.

When salts (or ionic compounds) dissolve in water, it is assumed that they dissociate entirely into separate ions.

Some of these ions react with water and behave as acids or bases.

The acidic or basic nature of a salt solution depends on whether it is a product of a:

1) strong acid-strong base reaction,

2) a weak acid-strong base reaction,

3) a strong acid-weak base reaction, or

4) a weak acid-weak base reaction.

Salts of Acid–Base Reactions

Salts

Salts are the products of acid-base reactions.

The general reaction to produce salt:

$$\text{Acid} + \text{Base} \rightarrow \text{Salt} + \text{Water}$$

The acid and base are neutralized; H^+ and OH^- form water.

The acid's nonmetallic ions and metal ions of the base form the salt.

For example, NaCl is a product of the following acid-base reaction:

$$HCl\ (aq) + NaOH\ (aq) \rightarrow NaCl\ (aq) + H_2O$$

In the chemical formula of salt, the cation is contributed by the base (e.g., Na^+), while the acid contributes the anion (e.g., Cl^-).

Four types of salts

The four types of salts are:

- **Salts of strong acid-strong base** reactions (e.g., NaCl, KNO_3, $NaClO_4$)
- **Salts of weak acid-strong base** reactions (e.g., $NaC_2H_3O_3$, K_2CO_3, KCN, $NaCHO_2$)
- **Salts of strong acid-weak base** reactions (e.g., NH_4Cl, NH_4NO_3, $HONH_3Cl$)
- **Salts of weak acid-weak base** reactions (e.g., $NH_4C_2H_3O_2$, NH_4CN, NH_4HS)

When dissolved in water, these salts produce acidic, basic, or neutral solutions.

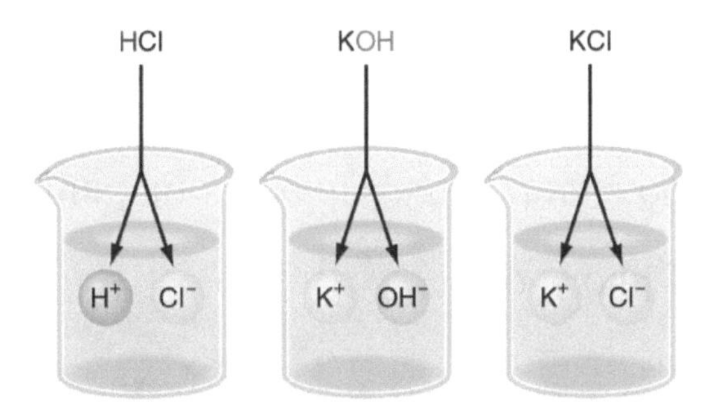

The acidic dissociation constant increases with stable anions

Salts of ***Strong Acid-Strong Base*** *Reactions*: (e.g., NaCl, $NaNO_3$, KBr)

- Salts form a neutral solution because neither the cation nor anion reacts with water and offsets the equilibrium concentrations of H_3O^+ and OH^- in the solution.

Salts of ***Weak Acid-Strong Base*** *Reactions*: (e.g., NaF, $NaNO_2$, $NaC_2H_3O_2$)

- Salts form basic solutions when dissolved in H_2O.

  The anions of such salts react with H_3O increasing [^-OH].

- Sodium acetate ($NaC_2H_3O_2$) is a product of the reaction between acetic acid ($HC_2H_3O_2$), which is a weak acid, and a strong base (NaOH).

$$HC_2H_3O_2\ (aq) + NaOH\ (aq) \rightarrow NaC_2H_3O_2\ (aq) + H_2O$$

Salts of ***Strong Acid-Weak Base*** *Reactions*: (e.g., NH_4Cl, NH_4NO_3, $(CH_3)_2NH_2Cl$, C_5H_5NHCl)

- Aqueous solutions of salts are acidic.
- The cations react with H_2O and increase [H_3O^+].
- Example: NH_4Cl is produced when HCl (strong acid) reacts with NH_3 (a weak base):

$$HCl\ (aq) + NH_3\ (aq) \rightarrow NH_4Cl\ (aq) \rightarrow NH_4^+\ (aq) + Cl^-\ (aq)$$

In an aqueous solution, NH_4^+ equilibrium increases [H_3O^+] and creates an acidic solution:

$$NH_4^+\ (aq) + H_2O \rightleftarrows H_3O^+\ (aq) + NH_3\ (aq)$$

$$K_a = [H_3O^+]\cdot[NH_3]\ /\ [NH_4^+]$$

While in an NH_3 solution, the following equilibrium occurs:

$$NH_3\ (aq) + H_2O \rightleftarrows NH_4^+\ (aq) + OH^-\ (aq)$$

$$K_b = [NH_4^+]\cdot[^-OH]\ /\ [NH_3]$$

$$K_a \times K_b = \{[H_3O^+]\cdot[NH_3]\ /\ [NH_4^+]\} \times \{[NH_4^+]\cdot[^-OH]\ /\ [NH_3]\}$$

$$K_a \times K_b = K_w = [H_3O^+]\cdot[^-OH]$$

$$K_w = 1.0 \times 10^{-14}$$

For NH_4^+ :

$K_a = K_w / K_b$ (for NH_3)

$K_a = (1.0 x 10^{-14}) / (1.8 x 10^{-5})$

$K_a = 5.6 \times 10^{-10}$

An aqueous solution of NH_4^+ has a $K_a = 5.6 \times 10^{-10}$ at 25 °C (which is > K_w).

A 0.10 *M* solution of NH_4Cl or NH_4NO_3 has $[H_3O^+] \approx 7.5 \times 10^{-6}$ M and pH ≈ 5.1

*Salts of **Weak Acid-Weak Base** Reactions*: (e.g., $NH_4C_2H_3O_2$, NH_4CN, NH_4NO_2)

Aqueous salts can be neutral, acidic, or basic, depending on the relative magnitude of the K_a of the weak acid and the K_b of the weak base.

- If $K_a \approx K_b$, salt forms an approximately neutral solution.

For example:

K_a of $HC_2H_3O_2 = 1.8 \times 10^{-5}$

K_b of $NH_3 = 1.8 \times 10^{-5}$

When $NH_4C_2H_3O_2$ dissolves in water and dissociates, the following equilibria exist:

$NH_4C_2H_3O_2\,(aq) \rightarrow NH_4^+\,(aq) + C_2H_3CO_2^-\,(aq)$

$NH_4^+\,(aq) + H_2O \rightleftarrows H_3O^+\,(aq) + NH_3\,(aq)$

$K_a = 5.6 \times 10^{-10}$

$C_2H_3O_2^-\,(aq) + H_2O \rightleftarrows HC_2H_3O_2\,(aq) + OH^-\,(aq)$

$K_b = 5.6 \times 10^{-10}$

K_a (for NH_4^+) = K_b (for $C_2H_3O_2^-$), at equilibrium $[H_3O^+] = [^-OH]$ and the $NH_4C_2H_3O_2$ solution is neutral.

If $\boldsymbol{K_a > K_b}$, the salt solution is acidic.

For NH_4NO_2:

K_a $(HNO_2) = 4.0 \times 10^{-4}$

K_b $(NH_3) = 1.8 \times 10^{-5}$

$K_a > K_b \rightarrow$ acidic solution because hydrolysis forms a solution with $[H_3O^+] > [^-OH]$.

If $\boldsymbol{K_a < K_b}$, the salt solution is basic.

For NH_4CN in solution,

K_a $(HCN) = 6.2 \times 10^{-10}$

K_b $(NH_3) = 1.8 \times 10^{-5}$

The following equilibria exist:

$NH_4CN\ (aq) \rightarrow NH_4^+\ (aq) + CN^-\ (aq)$

$NH_4^+\ (aq) + H_2O \rightleftarrows H_3O^+\ (aq) + NH_3\ (aq)$

$K_a = 5.6 \times 10^{-10}$

$CN^-\ (aq) + H_2O \rightleftarrows HCN\ (aq) + OH^-\ (aq)$

$K_b = 1.6 \times 10^{-5}$

$K_b\ (CN^-) > K_b\ (NH_4^+)$, at equilibrium $[^-OH] > [H_3O^+]$; an aqueous solution of NH_4CN is basic.

pH Calculations for Acid–Base Solutions

Calculating pH of strong acid and strong base solutions

Strong acids are assumed to ionize completely in an aqueous solution.

For monoprotic acids (i.e., acids with a single ionizable hydrogen) such as HCl and HNO_3, the [hydronium ion; H_3O^+] in solution is the same as the molar concentration of the acid:

$[H_3O^+] = [HX]$

For example, consider 0.10 M HCl (*aq*)

$[H_3O^+] = 0.10$ M

pH = −log(0.10)

pH = 1.00

A strong base such as NaOH has [$^-$OH] equal to the molar concentration of dissolved NaOH.

A solution of 0.10 M NaOH (*aq*) has:

[$^-$OH] = 0.10 M

pOH = −log[$^-$OH]

−log[$^-$OH] = −log(0.10)

pOH = 1.00

pH = 14.00 – 1.00

pH = 13.00

A strong base such as $Ba(OH)_2$ produces twice the concentration of $^-$OH as the molar concentration of $Ba(OH)_2$ in solution:

$Ba(OH)_2\,(aq) \rightarrow Ba^{2+}\,(aq) + 2\,OH^-\,(aq)$

[$^-$OH] = 2 × [$Ba(OH)_2$]

In a solution of 0.010 M $Ba(OH)_2$

[$^-$OH] = 0.020 M

pOH = 1.70, and pH = 12.30

Calculating pH of weak acid solutions

Unlike strong acids, weak acids do not ionize completely.

At equilibrium, $[H^+]$ is less than the concentration of the acid.

The concentration of H^+ in a weak acid solution depends on the initial acid concentration and the K_a of the acid.

Determine $[H^+]$ of a weak acid using the "ICE" table as follows:

For example, consider a 0.10 M acetic acid solution and its ionization products.

$$HC_3H_3O_2\,(aq) \rightleftarrows H^+\,(aq) + C_2H_3O_2^-\,(aq)$$

Initial [ ], M:	0.10	0.00	0.00
Change, Δ[ ], M:	$-x$	$+x$	$+x$
Equilibrium [ ], M :	$(0.10 - x)$	x	x

The acid ionization constant, K_a, is given by the expression:

$$K_a = [H_3O^+]\cdot[C_2H_3O_2^-] / [CH_3COOH]$$

$$K_a = x^2 / (0.010 - x)$$

$$K_a = 1.8 \times 10^{-5}$$

Since,

$$K_a << 0.10$$

approximate that $x << 0.10$, and $(0.10 - x) \sim 0.10$

$$K_a = x^2 / (0.10 - x)$$

$$K_a \approx x^2 / 0.10$$

$$K_a = 1.8 \times 10^{-5}$$

$$x^2 = 1.8\ x\ 10^{-6}$$

$$x = \sqrt{(1.8\ x\ 10^{-6})}$$

$$x = 1.3 \times 10^{-3}$$

continued…

Note that

$x = [H_3O^+]$

$[H_3O^+] = 1.3 \times 10^{-3}$ M

$pH = -\log(1.3 \times 10^{-3})$

pH = 2.89

Calculating pH of weak base solutions

The concentration of ^-OH in a weak base, such as NH_3 (*aq*), depends on its K_b value and the initial concentration of the base.

To determine $[^-OH]$ and pH of 0.10 M NH_3 (*aq*), set the following "ICE" table:

$NH_3\,(aq) + H_2O \rightleftarrows NH_4^+\,(aq) + {}^-OH\,(aq)$

Initial [ ], *M*:	0.10	0.00	0.00
Change, Δ[ ], *M*:	$-x$	$+x$	$+x$
Equilibrium [ ], *M*:	$(0.10 - x)$	x	x

$K_b = [NH_4^+]\cdot[^-OH] / [NH_3]$

$K_b = x^2 / (0.10 - x)$

$K_b = 1.8 \times 10^{-5}$

Using the approximation method

$K_b = x^2 / (0.10 - x)$

$K_b \approx x^2 / 0.10$

$K_b = 1.8 \times 10^{-5}$

$x^2 = 1.8 \times 10^{-6}$

$x = \sqrt{(1.8 \times 10^{-6})}$

$x = 1.3 \times 10^{-3}$

continued…

where

$x = [^{-}OH]$

$x = 1.3 \times 10^{-3}\ M$

$pOH = -\log(1.3 \times 10^{-3})$

$pOH = 2.87$

$pH = 11.13$

Calculating pH of basic or acidic salt solutions

1. For example, consider a solution of 0.050 *M* sodium acetate, which dissociates completely and establishes the following equilibrium:

   $NaC_2H_3O_2\,(aq) \rightarrow Na^+\,(aq) + C_2H_3O_2^-\,(aq)$

The acetate ion establishes the equilibrium in *aq* solution:

$C_2H_3O_2^-\,(aq) + H_2O \rightleftarrows HC_2H_3O_2\,(aq) + {}^{-}OH\,(aq)$

$K_b = [HC_2H_3O_2]\cdot[^{-}OH] / [C_2H_3O_2^-]$

$K_b = 5.6 \times 10^{-10}$

By approximation

$[^{-}OH] = \sqrt{}(K_b[C_2H_3O_2^-])$

$[^{-}OH] = \sqrt{}\{(5.6 \times 10^{-10})\cdot(0.050)\}$

$[^{-}OH] = 5.3 \times 10^{-6}$ M

$pOH = -\log(5.3 \times 10^{-6})$

$pOH = 5.28$

$pH = 8.72$ (solution is basic)

continued…

2. For example, consider a solution of 0.050 M NH_4Cl, which dissociates and establishes the following equilibrium:

$$NH_4Cl\ (aq) \rightarrow NH_4^+\ (aq) + Cl^-\ (aq)$$

$$NH_4^+\ (aq) + H_2O \rightleftarrows H_3O^+\ (aq) + NH_3\ (aq)$$

$$K_a = [H_3O^+]\cdot[NH_3] / [NH_4^+]$$

$$K_a = 5.6 \times 10^{-10}$$

By approximation

$$[H_3O^+] = \sqrt{}(K_a[NH_4^+])$$

$$[H_3O^+] = \sqrt{}\{(5.6 \times 10^{-10})\cdot(0.050)\}$$

$$[H_3O^+] = 5.3 \times 10^{-6}\ M$$

$$pH = -\log(5.3 \times 10^{-6})$$

$$pH = 5.28,\ \text{(solution is acidic)}$$

Notes for active learning

Buffered Solutions

Buffers

A *buffer* is a solution that maintains pH (little or no change) even when a small amount of strong acid or strong base is added. A buffer solution contains a weak acid and the "salt" of its conjugate base or a weak base and the "salt" of its conjugate acid.

The *buffering capacity* is the amount of H^+ or OH^- ion the buffer absorbs without significantly altering the pH.

A buffer with large concentrations of buffering components absorbs significant quantities of strong acid or strong base, with little change in its pH, has a large buffering capacity.

A buffer is effective within a pH range, typically about ±1 of the pK_a of its acid component.

Some *common buffer systems*:

Buffer	pK_a	pH Range
$HCHO_2 – NaCHO_2$	3.74	2.75–4.75
$HC_2H_3O_2 – NaC_2H_3O_2$	4.74	3.75–5.75
$KH_2PO_4 – K_2HPO_4$	7.21	6.20–8.20 (a blood buffer)
$CO_2/H_2O – NaHCO_3$	6.37	5.40–7.40 (a blood buffer)
$NH_3 – NH_4Cl$	9.25	8.25–10.25

When 0.01 mol of HCl is added to 1 L of pure water, $[H^+]$ increases from 10^{-7} to 10^{-2} M and the pH changes from about 7 to 2. This change in pH indicates that water is not a buffer.

When the same HCl amount is added to a solution containing a mixture of 1 M acetic acid ($HC_2H_3O_2$) and 1 M sodium acetate ($NaC_2H_3O_2$), the solution's pH changes little – it goes from 4.74 to 4.66.

A solution composed of acetic acid and sodium acetate is a buffer solution.

For example, consider a buffered solution composed of KH_2PO_4 and K_2HPO_4. The species present in the solution are primarily K^+, $H_2PO_4^-$ and HPO_4^{2-}. K^+ is a spectator ion and not involved in the buffering reaction.

If a small amount of strong acid is added to this solution, the H^+ ions from the acid react with the base component of the buffer (HPO_4^{2-}):

$$H^+ (aq) + HPO_4^{2-} (aq) \rightarrow H_2PO_4^- (aq) \ldots \text{(buffering reaction 1)}$$

When a strong base such as NaOH is added, the ^-OH reacts with the buffer's acid component ($HC_2H_3O_2$).

$$^-OH (aq) + HC_2H_3O_2 (aq) \rightarrow H_2O + C_2H_3O_2^- (aq) \ldots \text{(buffering reaction 2)}$$

Reactions (1) and (2) are critical buffering reactions that maintain the solution's pH.

Biological buffers

Buffered solutions are vital to living organisms. Metabolic reactions are controlled or accelerated by biological catalysts called *enzymes*, often proteins that function within a narrow pH range.

The human body fluid must be at a specific (narrow) pH range. Human blood is maintained at the pH range of 7.30 - 7.40. A drop below pH 7 or rise above pH 7.5 can be fatal.

Two buffer systems – the phosphate buffer ($H_2PO_4^-$-HPO_4^{2-}) and carbonic acid-bicarbonate buffer (H_2CO_3-HCO_3^-) are essential to maintaining the normal blood pH.

The buffering reactions of bicarbonate buffer are:

$$H^+ (aq) + HCO_3^- (aq) \rightarrow H_2O + CO_2 (aq)$$

$$^-OH (aq) + CO_2 (aq) + H_2O \rightarrow 2\ HCO_3^- (aq)$$

Buffers resist changes in pH

The *capacity of a buffer* is determined by the sizes of [AB] and [B^-].

Consider a buffer made up of acetic acid and sodium acetate, in which the significant species present in the solution are $HC_2H_3O_2$ and $C_2H_3O_2^-$.

For example, if a small amount of HCl (*aq*) is added to this solution, most H^+ (from HCl) is absorbed by the conjugate base, $C_2H_3O_2^-$.

$$H^+ (aq) + C_2H_3O_2^- (aq) \rightarrow HC_2H_3O_2^- (aq)$$

Since $C_2H_3O_2^-$ is present in a larger quantity than the added H^+, the reaction shifts almost entirely to the right. This buffering reaction prevents a significant increase in [H^+] and minimizes the pH change.

For example, if a strong base, such as NaOH (*aq*), is added, most ^{-}OH ions (from NaOH) react with the buffer's acidic component.

$$^{-}OH\ (aq) + HC_2H_3O_2\ (aq) \rightarrow H_2O + C_2H_3O_2^{-}\ (aq)$$

Because of the larger concentration of $HC_2H_3O_2$ compared to ^{-}OH, this reaction goes almost to completion.

This buffering reaction prevents a significant increase in $[^{-}OH]$ and minimizes pH changes.

For example, for a buffer containing the weak acid HB and the salt NaB, such that B^- is the conjugate base to the acid, the concentration $[H^+]$ and pH of the buffer depending on the dissociation constant, K_a, of the acid component, and the concentration ratio $[B^-] / [HB]$ in the buffer solution.

For example, consider the equilibrium:

$$HB\ (aq) \leftrightarrows H^+\ (aq) + B^-\ (aq)$$

$$K_a = [H^+]\cdot[B^-] / [HB]$$

Rearranging the expression:

$$[H^+] = K_a \times ([HB] / [B^-])$$

$$pH = pK_a + \log([B^-] / [HB])$$

Calculating pH of buffered solutions

The *pH of a buffered solution* is determined by the ratio $[B^-] / [HB]$.

The following example illustrates how to calculate the change in pH and the buffering capacity of a buffered solution after a strong acid is added.

Calculate the change in pH when 0.010 mol of HCl adds to 1.0 L of each of the buffers:

Buffer A: 1.0 M $HC_2H_3O_2$ + 1.0 M $NaC_2H_3O_2$

Buffer B: 0.020 M $HC_2H_3O_2$ + 0.020 M $NaC_2H_3O_2$

Henderson-Hasselbalch equation

The expression $pH = pK_a + \log([B^-] / [HB])$ is the *Henderson-Hasselbalch equation*, which is useful for calculating the pH of solutions when the K_a and the ratio $[B^-] / [HB]$ are known.

For buffers, the *Henderson-Hasselbalch equation* calculates pH:

$$pH = pK_a + \log([B^-] / [HB])$$

$$pH = pK_a + \log([C_2H_3O_2^-] / [HC_2H_3O_2])$$

$$pH = -\log(1.8 \times 10^{-5}) + \log(1)$$

$$pH = 4.74$$

For example, 0.010 mol HCl is added to Buffer A with the following reaction:

$$H^+ (aq) + C_2H_3O_2^- (aq) \rightarrow HC_2H_3O_2 (aq)$$

[ ] before reaction:	0.010 M	1.0 M	1.0 M
[ ] after reaction:	0	0.99 M	1.01 M

The new pH:

$$pH = pK_a + \log([B^-] / [HB])$$

$$pH = 4.74 + \log(0.99 / 1.01)$$

$$pH = 4.74 - 0.010$$

$$pH = 4.73 \text{ (pH is changed} \approx 0.21\%)$$

0.010 mol HCl is added to Buffer B with the following reaction:

$$H^+ (aq) + C_2H_3O_2^- (aq) \rightarrow HC_2H_3O_2 (aq)$$

[ ] before rxn:	0.010 M	0.020 M	0.020 M
[ ] after rxn:	0	0.010 M	0.030 M

The expression to calculate pH:

$$pH = pK_a + \log([B^-] / [HB])$$

$pH = 4.74 + \log(0.010 / 0.030)$

$pH = 4.74 - 0.48$

$pH = 4.26$ (pH decreases by 10%)

From above, Buffer A, which contains larger quantities of buffering components, has a higher buffering capacity than Buffer B.

For Buffer A to decrease its pH by 0.48 units (or 10%), it must absorb the equivalent of 0.50 mol of HCl.

Summary of buffer solutions

1. The solution contains a weak acid HX and its conjugate base X^-, or a weak base B and its conjugate acid BH^+ in appreciable amounts.

2. A buffer solution maintains its pH by absorbing H^+ or OH^- produced by a strong acid or strong base, so these ions do not accumulate.

3. The buffering reaction involves H^+ with the conjugate base $\mathbf{X^-}$ in the buffer, or the reaction of ^-OH with the acid component (**HX**) of the buffer:

   $H^+ (aq) + \mathbf{X^-} (aq) \rightarrow HX (aq)$

   $^-OH (aq) + \mathbf{HX} (aq) \rightarrow H_2O + X^-$

   These two reactions prevent a significant increase in $[H^+]$ or $[^-OH]$ in the solution.

4. The *buffering capacity* of a solution implies the amount of H^+ or ^-OH it absorbs without significantly changing its pH. This depends on the concentration of the weak acid and its conjugate base in the solution.

5. The *buffering range* of a solution depends on the pK_a of the buffer's acid component.

   A buffer is most effective when the pH range = $pK_a \pm 1$.

Notes for active learning

Ionic Equations and Neutralization

Molecular, ionic and net ionic equations

Examples of molecular, complete (or total) ionic, and net ionic equations for strong acid-strong base reactions:

Molecular:

$$HCl\ (aq) + NaOH\ (aq) \rightarrow H_2O\ (l) + NaCl\ (aq)$$

Complete ionic:

$$H^+\ (aq) + Cl^-\ (aq) + Na^+\ (aq) + OH^-\ (aq) \rightarrow H_2O\ (l) + Na^+\ (aq) + Cl^-\ (aq)$$

Net ionic:

$$H^+\ (aq) + OH^-\ (aq) \rightarrow H_2O\ (l)$$

Spectator ions:

Na^+ and Cl^-

Weak acids and weak bases ionize partially, and most weak acids and bases remain in the molecular form in solution.

Therefore, weak acids and bases should NOT be written in ionized forms when writing the ionic equations.

The three equations for the reaction between acetic acid (a weak acid) and sodium hydroxide (a strong base) are:

Molecular:

$$HC_2H_3O_2\ (aq) + NaOH\ (aq) \rightarrow H_2O\ (l) + NaC_2H_3O_2\ (aq)$$

Complete ionic:

$$HC_2H_3O_2\ (aq) + Na^+\ (aq) + OH^-\ (aq) \rightarrow H_2O\ (l) + Na^+\ (aq) + C_2H_3O_2^-\ (aq)$$

Net ionic:

$$HC_2H_3O_2\ (aq) + OH^-\ (aq) \rightarrow H_2O\ (l) + C_2H_3O_2^-\ (aq)$$

Spectator ion:

Na^+

Neutralization

Neutralization is a chemical reaction where an acid and a base react quantitatively, resulting in no excess of hydrogen (H^+) or hydroxide ions (OH^-) in the aqueous solution.

The products of acid-base reactions are salt and water.

$$HCl\ (aq) + NaOH\ (aq) \rightarrow H_2O\ (l) + NaCl\ (aq)$$

$$HclO_4\ (aq) + KOH\ (aq) \rightarrow H_2O\ (l) + KclO_4\ (aq)$$

$$HC_2H_3O_2\ (aq) + NaOH\ (aq) \rightarrow H_2O\ (l) + NaC_2H_3O_2\ (aq)$$

Note that the substances involved are subject to dissociation.

Stoichiometry of acid-base neutralization

For example, how many mL of 0.1725 M NaOH (*aq*) are needed to neutralize 25.00 mL of 0.2040 M HCl (*aq*)? According to the equation:

$$HCl\ (aq) + NaOH\ (aq) \rightarrow H_2O\ (l) + NaCl\ (aq)$$

Moles of NaOH needed = Moles of HCl present

$$(\text{Liters NaOH} \times \text{Molarity NaOH}) = (\text{Liters HCl} \times \text{Molarity HCl})$$

$$(\text{Liters of NaOH} \times 0.1725\ \text{mol/L}) = (0.02500\ \text{L} \times 0.2040\ \text{mol/L})$$

Divide each side by 0.1725 mol/L:

$$\text{Liters of NaOH} = \frac{0.02500\ \text{L} \times 0.2040\ \text{mol/L}}{0.1725\ \text{mol/L}}$$

$$\text{Liters of NaOH} = 0.02957\ \text{L} = 29.57\ \text{mL}$$

continued...

For example, if 10.00 mL of acetic acid $HC_2H_3O_2$ of an unknown concentration requires 38.64 mL of 0.2250 M KOH to neutralize, what is the molarity of the acetic acid?

From the reaction:

$$HC_2H_3O_2\,(aq) + NaOH\,(aq) \rightarrow H_2O\,(l) + NaC_2H_3O_2\,(aq)$$

Moles of acetic acid = Moles of NaOH

(Liter of acid × Molarity of acid) = (Liter of base × Molarity of base)

(0.01000 L × Molarity $HC_2H_3O_2$) = (0.03864 L × 0.2250 M NaOH)

Divide each side by 0.01000 L:

$$\text{Molarity of } HC_2H_3O_2 = \frac{0.03864\text{ L} \times 0.2250\text{ M}}{0.01000\text{ L}}$$

$HC_2H_3O_2$ = 0.8694 M

Notes for active learning

Titration

Molar concentration determination

Titration (or *volumetric analysis*) is a standard laboratory procedure to determine an *unknown concentration*.

Titration (*volumetric analysis*) is an essential application of neutralization reactions. A reagent of known concentration is slowly added to a sample of unknown concentration until the neutralization reaction (i.e., a reaction between an acid and a base) is complete.

Titration determines the molar concentration of a solution (analyte) using the volume and concentration of another (titrant) by adding an exact amount of the *titrant* from a buret to another reactant (*analyte*) in a flask.

The *analyte* is the substance (i.e., sample) of an unknown concentration being analyzed.

The *titrant* is the analytical reagent of known concentration added to the sample. It is carefully added from a buret until the *equivalence point* (i.e., the point of neutralization) is reached.

If the titrant's volume and concentration are known, its number of moles can be calculated.

An *indicator* is added to the reaction to determine the endpoint of a redox titration (i.e., when sample molecules have been depleted).

The *endpoint* is marked by an indicator's color change when a solution transitions from acidic to slightly basic. A redox indicator undergoes a color change at the neutralization reaction's equivalence point.

In a useful titration, the *equivalence point* should match the *endpoint.*

The number of moles of the analyte and its concentration can be determined from reaction stoichiometry.

Titration experimental factors

1. The reaction between titrant and analyte occurs rapidly.
2. The balanced equation is known.
3. The endpoint occurs precisely at or close to the equivalence point.
4. The volume of titrant to reach the equivalence point is accurately measurable.

Stoichiometry determines unknown concentrations

An acid (e.g., HCl (*aq*)) is titrated with aqueous NaOH as follows:

$$HCl\ (aq) + NaOH\ (aq) \rightarrow NaCl\ (aq) + H_2O$$

For example, the stoichiometric ratio of HCl to NaOH is 1 mole HCl to 1 mole NaOH. The standard solution's volume and concentration (acid or base) are known in titration.

However, only the volume of the other solution (acid or base), whose concentration is to be determined, is known.

The above stoichiometry enables calculating unknown concentrations.

For example, suppose that 25.00 mL of 0.2250 M HCl (*aq*) is required to neutralize 27.45 mL of aqueous NaOH solution, whose concentration is unknown. Calculate the moles of each reactant and the concentration of NaOH.

No. of mol of HCl reacted:

$$25.0\text{ mL} \times (1\text{ L}/1{,}000\text{ mL}) \times (0.2250\text{ mol/L}) = 0.005625\text{ mol}$$

Since HCl and NaOH react in a 1:1 ratio,

No. of mol of NaOH = No. of mol of HCl

No. of mol of NaOH = 0.005625 mol

Molarity of NaOH = (0.005625 mol / 0.02745 L)

Molarity of NaOH = 0.2049 M

The stoichiometric ratio may not be 1 to 1, as below.

For example, suppose 20.0 mL H_2SO_4 of unknown concentration requires 32.0 mL of 0.205 M NaOH. Calculate the concentration of the H_2SO_4 solution.

$$H_2SO_4\ (aq) + 2\ NaOH\ (aq) \rightarrow Na_2SO_4\ (aq) + 2\ H_2O$$

The stoichiometric ratio is 1 mole of H_2SO_4 to 2 moles of NaOH.

$$\text{No. of mol of NaOH reacted} = \frac{0.205\text{ mol NaOH}}{1\text{ L solution}} \times 32.0\text{ mL} \times \frac{1\text{ L}}{1{,}000\text{ mL}}$$

No. of mol of NaOH reacted = 0.00656 mol

No. of mol of H_2SO_4 = Mol NaOH × (1 mol H_2SO_4 / 2 mol NaOH)

continued…

No. of mol of H_2SO_4 = 0.00656 mol NaOH × (1 mol H_2SO_4 / 2 mol NaOH)

No. of mol of H_2SO_4 = 0.00328 mol

$$\text{Molarity of } H_2SO_4 = \frac{0.00328 \text{ mol } H_2SO_4}{0.0200 \text{ L}}$$

Molarity of H_2SO_4 = 0.164 M

Strong acid–strong base titrations

For example, consider the strong acid-strong base titration of 20.0 mL of 0.100 M HCl (*aq*) with 0.100 M NaOH (*aq*) solution. Calculate the pH of the acid solution:

(a) before NaOH is added,

(b) after 15.0 mL of NaOH is added,

(c) after 19.5 mL of NaOH is added,

(d) after 20.0 mL of NaOH is added,

(e) after 21.0 mL of NaOH is added and

(f) after 25.0 mL of NaOH is added.

(a) Before titration:

$[H^+]$ = 0.100 M

pH = 1.000

(b) When 15.0 mL of NaOH is added, the following reaction occurs:

$H^+ (aq) + OH^- (aq) \rightarrow H_2O$

[ ] before mixing:

0.100 M 0.100 M

[ ] after mixing, but before reaction:

0.0571 M 0.0429 M

[ ] after reaction:

0.0142 M 0 M

$[H^+] = 0.0142$ M

pH = −log(0.0142)

pH = 1.848

(c) After 19.5 mL of NaOH is added, the calculation of $[H^+]$ is:

$H^+ (aq) + {}^-OH (aq) \rightarrow H_2O$

[ ] before mixing:

0.100 M 0.100 M

[ ] after mixing, but before reaction:

0.0506 M 0.0494 M

[ ] after reaction:

0.0012 M 0 M

$[H^+] = 0.0012$ M

pH = −log(0.0012)

pH = 2.92

Before the equivalent point, $[H^+]$ can be calculated:

$$[H^+] = \frac{(\text{initial mol of } H^+ - \text{mol of } {}^-OH \text{ added})}{(\text{L of HCl titrated} + \text{L of NaOH added})}$$

$$[H^+] = \frac{(0.00200 \text{ mol } H^+ - 0.00195 \text{ mol } {}^-OH)}{(0.0200 \text{ L of HCl} + 0.0195 \text{ L NaOH})}$$

$[H^+] = (0.000050$ mol / 0.0395 L)

$[H^+] = 0.0013$ M

pH = 2.90

(d) When 20.0 mL of 0.100 M NaOH has been added,

$H^+ (aq) + {}^-OH (aq) \rightarrow H_2O$

[ ] after mixing, but before reaction:

0.0500 M 0.0500 M

[ ] after reaction:

0 M 0 M

This point of the titration is the equivalence point, whereby only Na^+ and Cl^- occur in the solution.

Since neither reacts with water, the solution has a pH = 7.00.

(e) When 21.0 mL of 0.100 M NaOH has been added, there is excess OH^-:

$H^+ (aq) + {}^-OH (aq) \rightarrow H_2O$

[ ] after mixing, but before reaction:

0.0488 M 0.0512 M

[ ] after reaction:

0 0.0024 M

$[{}^-OH] = 0.0024$ M

pOH = 2.62

pH = 11.38

(f) When 25.0 mL of NaOH is added,

$H^+ (aq) + {}^-OH (aq) \rightarrow H_2O$

[ ] after mixing, but before reaction:

0.0444 M 0.0556 M

[ ] after reaction:

0 0.0112 M

$[^{-}OH] = 0.0112$ M

pOH = 1.953

pH = 12.047

In strong acid-strong base titrations, an abrupt change from about pH 3 to 11 occurs within ±0.5 mL of NaOH, added near the equivalent point.

Weak acid–strong base titrations

An example of a weak acid-strong base titration follows.

General reaction:

$$HA\ (aq) + {}^{-}OH\ (aq) \rightarrow H_2O + A^-\ (aq)$$

When acetic acid (weak acid) is titrated with sodium hydroxide (strong acid), the net reaction is:

$$HC_2H_3O_2\ (aq) + {}^{-}OH\ (aq) \rightarrow C_2H_3O_2^-\ (aq) + H_2O$$

For example, consider the titration of 20.0 mL of 0.100 M $HC_2H_3O_2$ (*aq*) with 0.100 M NaOH (*aq*) solution.

Calculate the pH of the solution:

(a) before the NaOH is added;

(b) after 10.0 mL of NaOH is added;

(c) after 15.0 mL of NaOH is added;

(d) after 20.0 mL of NaOH is added;

(e) after 25.0 mL of NaOH is added.

(a) Before titration:

$[H^+] = \sqrt{}(0.100\text{ M})\cdot(1.8 \times 10^{-5})$

$[H^+] = 1.34 \times 10^{-3}$ M

$pH = -\log(1.34 \times 10^{-3})$

pH = 2.873

(b) After adding 10.0 mL of 0.100 M NaOH, $[H^+]$ is calculated as follows:

$HC_2H_3O_2\,(aq) + OH^-\,(aq) \rightarrow C_2H_3O_2^-\,(aq) + H_2O$

Before mixing:

0.100 M   0.1000 M   0.0000 M

After mixing (before rxn):

0.0667 M   0.0333 M   0.0000 M

After reaction:

0.0333 M   0.0000 M   0.0333 M

Using the Henderson-Hasselbalch equation,

$pH = pK_a + \log([B^-] / [HB])$

$$[H^+] = K_a \times \frac{[HC_2H_3O_2]}{[C_2H_3O_2^-]}$$

$[H^+] = 1.8 \times 10^{-5} \times (0.0333\ M / 0.0333\ M)$

$[H^+] = 1.8 \times 10^{-5}\ M$

$pH = pK_a + \log([C_2H_3O_2^-] / [HC_2H_3O_2])$

$pH = 4.74 + \log(1)$

$pH = 4.74$

When a weak acid is half-neutralized, 50% of the acid is converted to its conjugate base.

Halfway to the equivalence point,

$[C_2H_3O_2^-] = [HC_2H_3O_2]$

Under this condition,

$[H^+] = K_a$, and $pH = pK_a$

(c) After adding 15.0 mL of 0.100 M NaOH, $[H^+]$ is calculated as follows:

$HC_2H_3O_2\,(aq) + {}^{-}OH\,(aq) \rightarrow C_2H_3O_2^-\,(aq) + H_2O$

Before mixing:	0.100 M	0.100 M	0.000 M
After mixing (before rxn):	0.0571 M	0.0429 M	0.000
After reaction:	0.0142 M	0.000 M	0.0429 M

Using the Henderson-Hasselbalch equation:

$pH = pK_a + \log([B^-] / [HB])$

$[H^+] = K_a \times [HC_2H_3O_2] / [C_2H_3O_2^-]$

$[H^+] = 1.8 \times 10^{-5} \times (0.0142\ M / 0.0429\ M)$

$[H^+] = 6.0 \times 10^{-6}\ M$

$pH = pK_a + \log([C_2H_3O_2^-] / [HC_2H_3O_2])$

$pH = 4.74 + \log(3.02)$

$pH = 4.74 + 0.48$

$pH = 5.22$

(d) At the equivalent point, when 20.0 mL of 0.100 M NaOH has been added, acid has been reacted and converted to its conjugate base.

The latter undergoes hydrolysis (reacts with water) as follows:

$C_2H_3O_2^-\,(aq) + H_2O\,(l) \leftrightarrows HC_2H_3O_2\,(aq) + {}^{-}OH\,(aq)$

Initial [ ], M:

0.0500	0.000	0.000

Change, Δ[ ], M:

$-x$	$+x$	$+x$

Equilibrium [ ], M:

$(0.0500 - x)$	x	x

By approximation:

$[^-OH] = x = \sqrt{(K_b \times [C_2H_3O_2^-]_0)}$

$[^-OH] = \sqrt{(5.6 \times 10^{-10}) \cdot (0.0500)}$

$[^-OH] = 5.3 \times 10^{-6}$ M

$pOH = -\log(5.3 \times 10^{-6})$

$pOH = 5.28$

$pH = 8.72$

Since the conjugate base of a weak acid undergoes hydrolysis (reacts with water), the pH of the solution at the equivalence point is greater than 7.00.

In the case of acetic-NaOH titration, the pH at the equivalence point is about 8.72.

For weak acids with a larger K_a, the pH at the equivalence point is closer to neutral pH; for a smaller K_a, the pH at the equivalence point is higher than neutral pH.

Notes for active learning

Indicators

Titration endpoint

The *titrant* is a solution of known concentration that is added (titrated) to another solution to determine the concentration of a second chemical species (i.e., analyte).

The *analyte* is the substance whose quantity or concentration is to be determined.

An *indicator* is a substance that changes color to mark a titration's endpoint. Indicators exhibit one color in acid (or protonated form, HIn) and another in base (or deprotonated form, In^-). Most indicators used in acid-base titration are weak organic acids.

Each indicator has a range of pH = pK_a ± 1, where the change of colors occurs.

A suitable indicator gives the *endpoint* corresponding to the titration's equivalence point. A pH range falls within the sharp increase (or decrease) of pH changes in the titration curves. The endpoint approximates the equivalence point with the known concentration of the titrant to calculate the amount or concentration of the analyte (i.e., unknown quantity or concentration).

Indicator equilibrium

Like a weak acid, an indicator has the following equilibrium in aqueous solution:

$$\text{HIn}\ (aq) \leftrightarrows \text{H}^+\ (aq) + \text{In}^-\ (aq)$$

$$K_a = \frac{[\text{H}^+]\cdot[\text{In}^-]}{[\text{HIn}]}$$

Rearranging

$$[\text{H}^+] = K_a \times ([\text{HIn}] / [\text{In}^-])$$

$$\text{pH} = \text{p}K_a + \log([\text{In}^-] / [\text{HIn}]$$

when

$$[\text{HIn}] = 10 \times [\text{In}^-]$$

$$\text{pH} = \text{p}K_a + \log([\text{In}^-] / 10[\text{In}^-])$$

$\text{pH} = \text{pK}_a - 1.0$; the indicator assumes the color of the acid form.

when

$$[\text{In}^-] = 10 \times [\text{HIn}]$$

$$\text{pH} = \text{p}K_a + \log(10[\text{HIn}] / [\text{HIn}])$$

$\text{pH} = \text{pKa} + 1.0$; the indicator assumes the color of base form.

Common acid–base indicator

Phenolphthalein, the most common acid-base indicator, has $K_a \sim 10^{-9}$. Its acid form (HIn) is colorless, and the conjugate base form (In^-) is pink.

It is colorless when the solution's pH ≤ 8 when 90% or more of the species are in the acid form (HIn) and pink at pH ≥ 10 when 90% or more of the species are in the conjugate base form (In^-).

The pH range at which an indicator changes depends on the K_a.

For phenolphthalein, which has $K_a \sim 10^{-9}$, its color changes in the pH range 8–10.

It is a suitable indicator for strong acid-strong base titrations and weak acid-strong base titrations.

pH of common indicators

Indicators	Acid color	Base color	pH Range	Type of Titrations
Methyl orange	orange	yellow	3.2–4.5	strong acid-strong base strong acid-weak base
Bromocresol green	yellow	blue	3.8–5.4	strong acid-strong strong acid-weak base
Methyl red	red	yellow	4.5–6.0	strong acid-strong base strong acid-weak base
Bromothymol blue	yellow	blue	6.0–7.6	strong acid-strong base
Phenol Red	orange	red	6.8–8.2	strong acid-strong base weak acid-strong base

Titration Curves

Interpreting titration curves

A *pH curve* is a graph of the pH of the solution *vs*. the volume of titrant in an acid-base titration.

The *equivalence point* is the point in a titration at which the amount of titrant added is just enough to neutralize the analyte solution completely.

At the equivalence point in an acid-base titration, moles of base = moles of acid, and the solution only contains salt and water.

The *buffering region* pH is usually $pK_a \pm 1$ (or $14 - pK_b \pm 1$), which resists pH changes before the curve's smooth (horizontal) portion is reached.

The data used to plot a pH curve may be obtained by computation *or* measuring a pH directly with the pH meter during titration.

The acid is incrementally added to the alkali (how titrations usually are performed).

Strong acid–strong base titration

The following titration curve is for the reaction:

$$NaOH\ (aq) + HCl\ (aq) \rightarrow NaCl\ (aq) + H_2O\ (l)$$

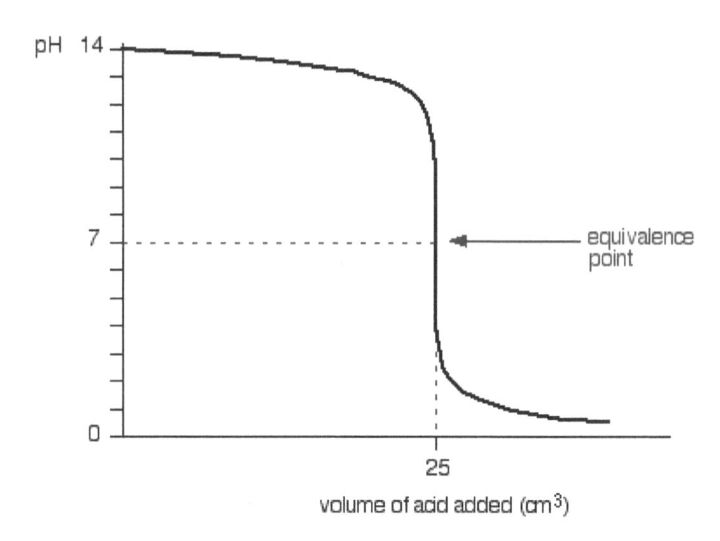

The pH falls a minimal amount until near the equivalence point, then a steep drop.

After the *equivalence point*, it is like the end of the strong acid-strong base reaction.

The buffering region's pH is usually $pK_a \pm 1$ (or $14 - pK_b \pm 1$), which resists pH changes before the curve's smooth (horizontal) portion is reached.

Strong acid–weak base titration

The following titration curve is for the reaction:

NH_3 (*aq*) + HCl (*aq*) → NH_4Cl (*aq*)

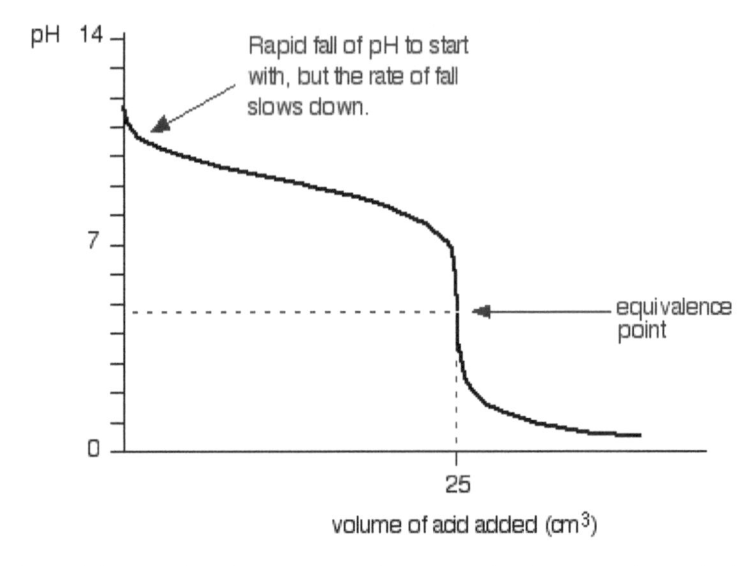

Since it is a weak base rather than a strong base, the titration curve's beginning is vastly different.

The pH decreases steeply as the acid is added, but the curve soon becomes less steep because a buffer solution is created composed of excess ammonia and formed ammonium chloride.

The equivalence point is now acidic (pH ~ 5) but is on the curve's steepest part.

The *equivalence point* is the point in a titration at which the amount of titrant added is just enough to neutralize the analyte solution completely.

At the equivalence point in an acid-base titration, moles of base = moles of acid, and the solution only contains salt and water.

After the *equivalence point*, it is like the end of the strong acid-weak base reaction.

The buffering region's pH is usually $pK_a \pm 1$ (or $14 - pK_b \pm 1$), which resists pH changes before the curve's smooth (horizontal) portion is reached.

Weak acid–strong base titration

The following titration curve is for the reaction:

CH_3COOH (*aq*) + NaOH (*aq*) → CH_3COONa (*aq*) + H_2O (*aq*)

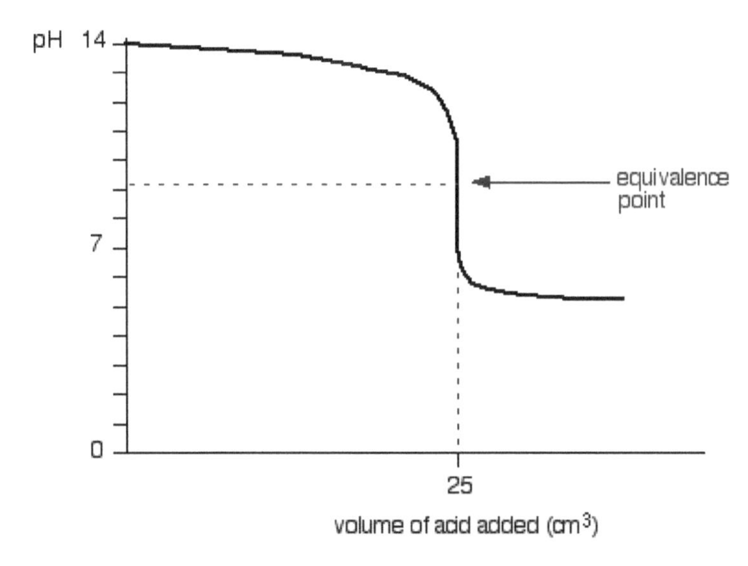

In the first part of the graph, there is an excess of sodium hydroxide, and this part of the curve is like the strong acid-strong base titration.

The *equivalence point* is the point in a titration at which the amount of titrant added is just enough to neutralize the analyte solution completely.

At the equivalence point in an acid-base titration, moles of base = moles of acid, and the solution only contains salt and water.

After the *equivalence point*, it is like the end of the weak acid-strong base reaction.

Once the acid is in excess, there is a difference between forming a buffer solution containing sodium ethanoate and ethanoic acid.

The buffering region's pH is usually $pK_a \pm 1$ (or $14 - pK_b \pm 1$), which resists pH changes before the curve's smooth (horizontal) portion is reached.

Weak acid–weak base titration

The following titration curve is for the reaction:

CH_3COOH (*aq*) + NH_3 (*aq*) → CH_3COONH_4 (*aq*)

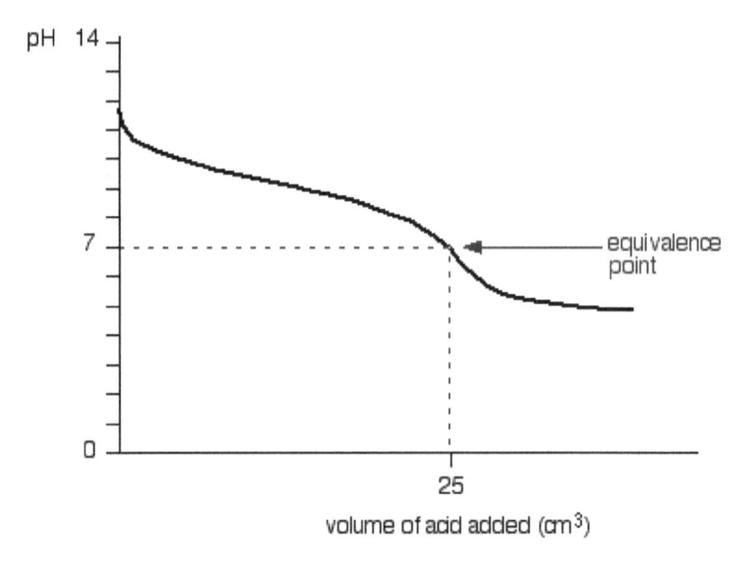

The acid and the base are equally weak, so the equivalence point is around pH 7.

This titration curve is essentially a combination of the two previous graphs.

The *equivalence point* is the point in a titration at which the amount of titrant added is just enough to neutralize the analyte solution completely.

At the equivalence point in an acid-base titration, moles of base = moles of acid, and the solution only contains salt and water.

After the *equivalence point*, it is like the end of the weak acid-weak base reaction.

There is no steep portion of this graph; the lack of a steep portion is an important identifying factor of a weak acid-weak base titration curve.

The buffering region's pH is usually $pK_a \pm 1$ (or $14 - pK_b \pm 1$), which resists pH changes before the curve's smooth (horizontal) portion is reached.

Four titration curves comparison

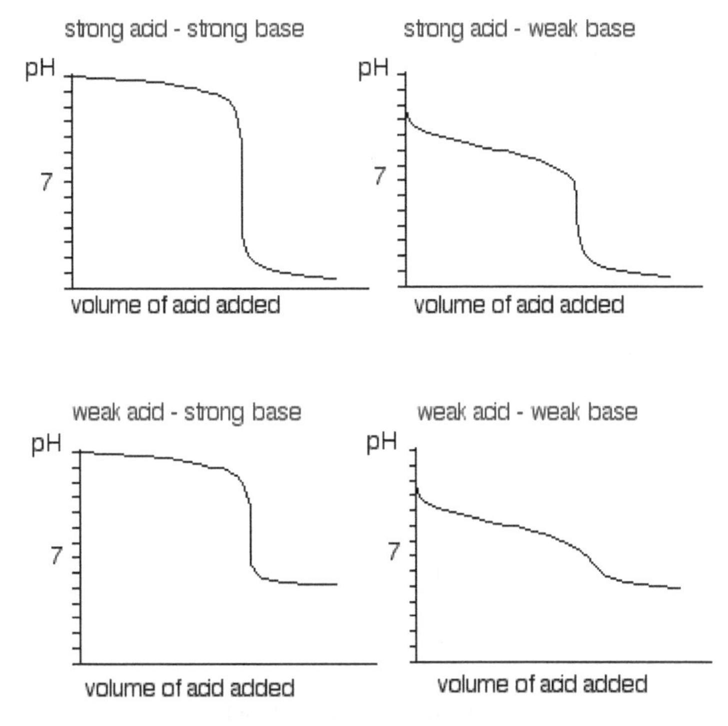

Redox titration transfers electrons

A *redox titration* is a type of titration based on the transfer of electrons.

In redox titration, standard solutions of oxidizing agents are used because solutions of reducing agents may react with oxygen in the air.

The analyzed species must be in a single oxidation state before titration.

For example, when iron ore is dissolved in hydrochloric acid, Fe^{2+} and Fe^{3+}, as iron(II) and iron(III), are present.

However, since it must be in a single oxidation state before titration with a standard $KMnO_4$ solution, iron(III) is reduced to iron(II) by reaction with excess zinc:

$$2\ Fe^{3+}(aq) + Zn\ (s) \rightarrow 2\ Fe^{2+}(aq) + Zn^{2+}(aq)$$

$$MnO_4^-(aq) + 5\ Fe^{2+}(aq) + 8\ H^+(aq) \rightarrow Mn^{2+}(aq) + 5\ Fe^{3+}(aq) + 4\ H_2O\ (l)$$

$KMnO_4$ and $K_2Cr_2O_7$ are the most used oxidizing agents in redox titration because the solutions change color during reduction; this color change serves as the titration indicator.

$KMnO_4$ is a purple solution but becomes colorless when MnO_4^- is reduced to Mn^{2+}.

The bright orange of $K_2Cr_2O_7$ changes to purplish-blue when $Cr_2O_7^{2-}$ is reduced to Cr^{3+}.

Sample redox titration

For example, a solution contains iron(II) and iron(III) ions. A 50.0 mL solution sample is titrated with 35.0 mL of 0.00280 M $KMnO_4$, which oxidizes Fe^{2+} to Fe^{3+}. The permanganate ion is reduced to manganese(II) ion. Another 50.0 mL solution sample is treated with zinc metal, which reduces Fe^{3+} to Fe^{2+}. The resulting solution is again titrated with 0.00280 M $KMnO_4$ and 48.0 mL is required. What are the concentrations of Fe^{2+} and Fe^{3+} in the solution?

For this problem, several steps are required.

1) The stoichiometric relationship of Fe(II) to permanganate is five to one:

$$MnO_4^- \,(aq) + 5\,Fe^{2+}\,(aq) + 8\,H^+\,(aq) \rightarrow Mn^{2+}\,(aq) + 5\,Fe^{3+}\,(aq) + 4\,H_2O\,(l)$$

2) Calculate the moles of Fe(II) reacted:

$$(0.00280\text{ mol/L})\,(0.0350\text{ L}) = 0.000098\text{ mol of }MnO_4^-$$

$$(0.0000980\text{ mol Mn})\cdot(5\text{ mol Fe} / 1\text{ mol Mn}) = 0.000490\text{ mol Fe(II)}$$

3) Determine the total iron content:

$$(0.00280\text{ mol/L})\cdot(0.0480\text{ L}) = 0.0001344\text{ mol of }MnO_4^-$$

$$(0.0000980\text{ mol Mn})\cdot(5\text{ mol Fe} / 1\text{ mol Mn}) = 0.000672\text{ mol total Fe}$$

4) Determine Fe(III) in solution and its molarity:

$$0.000672\text{ mol} - 0.000490\text{ mol} = 0.000182\text{ mol}$$

$$(0.000182\text{ mol}) / (0.050\text{ L}) = 0.00364\text{ M Fe(III)}$$

5) Determine the molarity of Fe(II):

$$(0.000490\text{ mol}) / (0.050\text{ L}) = 0.0098\text{ M Fe(II)}$$

Notes for active learning

Notes for active learning

PRACTICE QUESTIONS
&
DETAILED EXPLANATIONS

Practice Questions

==

Practice Set 1: Questions 1–20

==

1. Which is the conjugate acid–base pair in the reaction?

$CH_3NH_2 + HCl \leftrightarrow CH_3NH_3^+ + Cl^-$

A. HCl and Cl^-
B. $CH_3NH_3^+$ and Cl^-
C. CH_3NH_2 and Cl^-
D. CH_3NH_2 and HCl
E. HCl and H_3O^+

2. What is the pH of an aqueous solution if the $[H^+] = 0.10$ M?

A. 0.0
B. 1.0
C. 2.0
D. 10.0
E. 13.0

3. Which of the following pairs of substances are both species salts?

A. NH_4F and KCl
B. $CaCl_2$ and HCN
C. $LiOH$ and K_2CO_3
D. $NaOH$ and $CaCl_2$
E. HCN and K_2CO_3

4. Which reactant is a Brønsted-Lowry acid?

$HCl\ (aq) + KHS\ (aq) \rightarrow KCl\ (aq) + H_2S\ (aq)$

A. KCl
B. H_2S
C. HCl
D. KHS
E. None of the above

5. If a light bulb in a conductivity apparatus glows brightly when testing a solution, which of the following must be true about the solution?

A. It is highly reactive
B. It is slightly reactive
C. It is highly ionized
D. It is slightly ionized
E. It is not an electrolyte

6. What is the term for a substance capable of either donating or accepting a proton in an acid–base reaction?

I. Nonprotic II. Aprotic III. Amphoteric

A. I only
B. II only
C. III only
D. I and III only
E. II and III only

7. Which of the following compounds is a strong acid?

I. $HClO_4$ (*aq*) II. H_2SO_4 (*aq*) III. HNO_3 (*aq*)

A. I only
B. II only
C. II and III only
D. I and II only
E. I, II and III

8. What is the approximate pH of a strong acid solution where $[H_3O^+] = 8.30 \times 10^{-5}$?

A. 9 **B.** 11 **C.** 3 **D.** 5 **E.** 4

9. Which of the following reactions represents the ionization of H_2O?

A. $H_2O + H_2O \rightarrow 2\ H_2 + O_2$
B. $H_2O + H_2O \rightarrow H_3O^+ + {}^-OH$
C. $H_2O + H_3O^+ \rightarrow H_3O^+ + H_2O$
D. $H_3O^+ + {}^-OH \rightarrow H_2O + H_2O$
E. None of the above

10. Which set below contains only weak electrolytes?

A. NH_4Cl (*aq*), $HClO_2$ (*aq*), HCN (*aq*)
B. NH_3 (*aq*), HCO_3^- (*aq*), HCN (*aq*)
C. KOH (*aq*), H_3PO_4 (*aq*), $NaClO_4$ (*aq*)
D. HNO_3 (*aq*), H_2SO_4 (*aq*), HCN (*aq*)
E. NaOH (*aq*), H_2SO_4 (*aq*), $HC_2H_3O_2$ (*aq*)

11. Which of the following statements describes a neutral solution?

A. $[H_3O^+] / [{}^-OH] = 1 \times 10^{-14}$
B. $[H_3O^+] / [{}^-OH] = 1$
C. $[H_3O^+] < [{}^-OH]$
D. $[H_3O^+] > [{}^-OH]$
E. $[H_3O^+] \times [{}^-OH] \neq 1 \times 10^{-14}$

12. Which of the following is an example of an Arrhenius acid?

A. H_2O (*l*)
B. RbOH (*aq*)
C. $Ba(OH)_2$ (*aq*)
D. $Al(OH)_3$ (*s*)
E. None of the above

13. In the following reaction, which reactant is a Brønsted-Lowry base?

$$H_2CO_3\ (aq) + Na_2HPO_4\ (aq) \rightarrow NaHCO_3\ (aq) + NaH_2PO_4\ (aq)$$

A. $NaHCO_3$
B. NaH_2PO_4
C. Na_2HPO_4
D. H_2CO_3
E. None of the above

14. Which of the following is the conjugate base of ^-OH?

A. O_2
B. O^{2-}
C. H_2O
D. O^-
E. H_3O^+

15. What of the following describes the solution for a vinegar sample at pH of 5?

A. Weakly basic
B. Neutral
C. Weakly acidic
D. Strongly acidic
E. Strongly basic

16. What is the pI for glutamic acid that contains two carboxylic acid groups and an amino group? (Use the carboxyl $pK_{a1} = 2.2$, carboxyl $pK_{a2} = 4.2$ and amino $pK_a = 9.7$)

A. 3.2
B. 1.0
C. 6.4
D. 5.4
E. 5.95

17. Which of the following compounds cannot act as an acid?

A. NH_3
B. H_2SO_4
C. HSO_4^{1-}
D. SO_4^{2-}
E. CH_3CO_2H

18. A weak acid is titrated with a strong base. When the concentration of the conjugate base is equal to the concentration of the acid, the titration is at the:

A. endpoint
B. equivalence point
C. buffering region
D. diprotic point
E. indicator zone

19. If $[H_3O^+]$ in an aqueous solution is 7.5×10^{-9} M, what is the $[^-OH]$?

A. 6.4×10^{-5} M
B. $3.8 \times 10^{+8}$ M
C. 7.5×10^{-23} M
D. 1.3×10^{-6} M
E. 9.0×10^{-9} M

20. Which species has a K_a of 5.7×10^{-10} if NH_3 has a K_b of 1.8×10^{-5}?

A. H^+
B. NH_2^-
C. NH_4^+
D. H_2O
E. $NaNH_2$

Practice Set 2: Questions 21–40

---

21. Which of the following are the conjugate bases of HSO_4^-, CH_3OH and H_3O^+, respectively:

A. SO_4^{2-}, CH_2OH^- and ^-OH
B. CH_3O^-, SO_4^{2-} and H_2O
C. SO_4^-, CH_3O^- and ^-OH
D. SO_4^-, CH_2OH^- and H_2O
E. SO_4^{2-}, CH_3O^- and H_2O

22. If 30.0 mL of 0.10 M $Ca(OH)_2$ is titrated with 0.20 M HNO_3, what volume of nitric acid is required to neutralize the base according to the following expression?

$$2\ HNO_3\ (aq) + Ca(OH)_2\ (aq) \rightarrow 2\ Ca(NO_3)_2\ (aq) + 2\ H_2O\ (l)$$

A. 30.0 mL
B. 15.0 mL
C. 10.0 mL
D. 20.0 mL
E. 45.0 mL

23. Which of the following expressions describes an acidic solution?

A. $[H_3O^+] / [^-OH] = 1 \times 10^{-14}$
B. $[H_3O^+] \times [^-OH] \neq 1 \times 10^{-14}$
C. $[H_3O^+] < [^-OH]$
D. $[H_3O^+] > [^-OH]$
E. $[H_3O^+] / [^-OH] = 1$

24. Which is incorrectly classified as an acid, a base, or an amphoteric species?

A. LiOH / base
B. H_2O / amphoteric
C. H_2S / acid
D. NH_4^+ / base
E. None of the above

25. Which of the following is the strongest weak acid?

A. CH_3COOH; $K_a = 1.8 \times 10^{-5}$
B. HF; $K_a = 6.5 \times 10^{-4}$
C. HCN; $K_a = 6.3 \times 10^{-10}$
D. HClO; $K_a = 3.0 \times 10^{-8}$
E. HNO_2; $K_a = 4.5 \times 10^{-4}$

26. What are the products from the complete neutralization of phosphoric acid with aqueous lithium hydroxide?

A. $LiHPO_4$ (*aq*) and H_2O (*l*)
B. Li_3PO_4 (*aq*) and H_2O (*l*)
C. Li_2HPO4 (*aq*) and H_2O (*l*)
D. LiH_2PO_4 (*aq*) and H_2O (*l*)
E. Li_2PO_4 (*aq*) and H_2O (*l*)

27. Which of the following compounds is NOT a strong base?

A. $Ca(OH)_2$ **B.** $Fe(OH)_3$ **C.** KOH **D.** NaOH **E.** $^-NH_2$

28. What is the $[H^+]$ in stomach acid that registers a pH of 2.0 on a strip of pH paper?

A. 0.2 M **B.** 0.1 M **C.** 0.02 M **D.** 0.01 M **E.** 2 M

29. Which statement is true about distinguishing between dissociation and ionization?

A. Ionization is the separation of existing charged particles
B. Dissociation produces new charged particles
C. Ionization involves polar covalent compounds
D. Dissociation involves polar covalent compounds
E. Ionization is reversible, while dissociation is irreversible

30. Which of the following is a general property of a basic solution?

I. Turns litmus paper red
II. Tastes sour
III. Causes the skin of the fingers to feel slippery

A. I only
B. II only
C. III only
D. I and II only
E. I, II and III

31. Which of the following compound-classification pairs is incorrectly matched?

A. HF – weak acid
B. $LiC_2H_3O_2$ –salt
C. NH_3 – weak base
D. HI – strong acid
E. $Ca(OH)_2$ – weak base

32. What is the term for a substance that releases H^+ in H_2O?

A. Brønsted-Lowry acid
B. Brønsted-Lowry base
C. Arrhenius acid
D. Arrhenius base
E. Lewis acid

33. Which molecule is acting as a base in the following reaction?

$^-OH + NH_4^+ \rightarrow H_2O + NH_3$

A. ^-OH
B. NH_4^+
C. H_2O
D. NH_3
E. H_3O^+

34. Citric acid is a triprotic acid with three carboxylic acid groups having pK_a values of 3.2, 4.8 and 6.4. At a pH of 5.7, what is the predominant protonation state of citric acid?

A. Three carboxylic acid groups are deprotonated
B. Three carboxylic acid groups are protonated
C. One carboxylic acid group is deprotonated, while two are protonated
D. Two carboxylic acid groups are deprotonated, while one is protonated
E. The protonation state cannot be determined

35. When fully neutralized by treatment with barium hydroxide, a phosphoric acid yields $Ba_2P_2O_7$ as one of its products. The parent acid for the anion in this compound is:

A. monoprotic acid
B. diprotic acid
C. triprotic acid
D. hexaprotic acid
E. tetraprotic acid

36. Does a solution become more or less acidic when a weak acid solution is added to a concentrated solution of HCl?

A. Less acidic because the concentration of OH^- increases
B. No change in acidity because [HCl] is too high to be changed by the weak solution
C. Less acidic because the solution becomes more dilute with a less concentrated solution of H_3O^+ being added
D. More acidic because more H_3O^+ is being added to the solution
E. More acidic because the solution becomes more dilute with a less concentrated solution of H_3O^+ being added

37. Which of the following is a triprotic acid?

A. HNO_3
B. H_3PO_4
C. H_2SO_3
D. $HC_2H_3O_2$
E. CH_2COOH

38. For which of the following pairs of substances do the two members of the pair NOT react?

A. Na_3PO_4 and HCl
B. KCl and NaI
C. HF and LiOH
D. $PbCl_2$ and H_2SO_4
E. All react to form products

39. What happens to the pH when sodium acetate is added to a solution of acetic acid?

A. Decreases due to the common ion effect
B. Increases due to the common ion effect
C. Remains constant because sodium acetate is a buffer
D. Remains constant because sodium acetate is neither acidic nor basic
E. Remains constant due to the common ion effect

40. Which of the following is the acidic anhydride of phosphoric acid (H_3PO_4)?

A. P_2O
B. P_2O_3
C. PO_3
D. PO_2
E. P_4O_{10}

==

Practice Set 3: Questions 41–60

==

41. Which of the following does NOT act as a Brønsted-Lowry acid?

A. CO_3^{2-}
B. HS^-
C. HSO_4^-
D. H_2O
E. H_2SO_4

42. Which of the following is the strongest weak acid?

A. HF; pK_a = 3.17
B. HCO_3^-; pK_a = 10.32
C. $H_2PO_4^-$; pK_a = 7.18
D. NH_4^+; pK_a = 9.20
E. $HC_2H_3O_2$; pK_a = 4.76

43. Why does boiler scale form on the walls of hot water pipes from groundwater?

A. Transformation of $H_2PO_4^-$ ions to PO_4^{3-} ions, which precipitate with the "hardness ions," Ca^{2+}, Mg^{2+}, Fe^{2+}/Fe^{3+}
B. Transformation of HSO_4^- ions to SO_4^{2-} ions, which precipitate with the "hardness ions," Ca^{2+}, Mg^{2+}, Fe^{2+}/Fe^{3+}
C. Transformation of HSO_3^- ions to SO_3^{2-} ions, which precipitate with the "hardness ions," Ca^{2+}, Mg^{2+}, Fe^{2+}/Fe^{3+}
D. Transformation of HCO_3^- ions to CO_3^{2-} ions, which precipitate with the "hardness ions," Ca^{2+}, Mg^{2+}, Fe^{2+}/Fe^{3+}
E. The reaction of the CO_3^{2-} ions present in groundwater with the "hardness ions,"Ca^{2+}, Mg^{2+}, Fe^{2+}/Fe^{3+}

44. Which of the following substances, when added to a solution of sulfoxylic acid (H_2SO_2), could be used to prepare a buffer solution?

A. H_2O
B. $HC_2H_3O_2$
C. KCl
D. HCl
E. $NaHSO_2$

45. Which of the following statements is NOT correct?

A. Acidic salts are formed by partial neutralization of a diprotic acid by a diprotic base
B. Acidic salts are formed by partial neutralization of a triprotic acid by a diprotic base
C. Acidic salts are formed by partial neutralization of a monoprotic acid by a monoprotic base
D. Acidic salts are formed by partial neutralization of a diprotic acid by a monoprotic base
E. Acidic salts are formed by partial neutralization of a polyprotic acid by a monoprotic base

46. Which of the following is the acid anhydride for $HClO_4$?

A. ClO **B.** ClO_2 **C.** ClO_3 **D.** ClO_4 **E.** Cl_2O_7

47. Identify the acid/base behavior of each substance for the reaction:

$H_3O^+ + Cl^- \rightleftharpoons H_2O + HCl$

A. H_3O^+ acts as an acid, Cl^- acts as a base, H_2O acts as a base, and HCl acts as an acid
B. H_3O^+ acts as a base, Cl^- acts as an acid, H_2O acts as a base, and HCl acts as an acid
C. H_3O^+ acts as an acid, Cl^- acts as a base, H_2O acts as an acid, and HCl acts as a base
D. H_3O^+ acts as a base, Cl^- acts as an acid, H_2O acts as an acid, and HCl acts as a base
E. H_3O^+ acts as an acid, Cl^- acts as a base, H_2O acts as a base, and HCl acts as a base

48. Given the pK_a values for phosphoric acid of 2.15, 6.87, and 12.35, what is the ratio of HPO_4^{2-} / $H_2PO_4^-$ in a typical muscle cell when the pH is 7.35?

A. 6.32×10^{-6}
B. 1.18×10^5
C. 0.46
D. 3.02
E. 3.31×10^3

49. If a light bulb in a conductivity apparatus glows dimly when testing a solution, which of the following must be true about the solution?

I. It is slightly reactive
II. It is slightly ionized
III. It is highly ionized

A. I only
B. II only
C. III only
D. I and II only
E. II and III only

50. Which of the following properties is NOT characteristic of an acid?

A. It is neutralized by a base
B. It has a slippery feel
C. It produces H^+ in water
D. It tastes sour
E. Its pH reading is less than 7

51. What is the term for a substance that releases hydroxide ions in water?

A. Brønsted-Lowry base
B. Brønsted-Lowry acid
C. Arrhenius base
D. Arrhenius acid
E. Lewis base

52. For the reaction below, which of the following is the conjugate acid of C_5H_5N?

$$C_5H_5N + H_2CO_3 \leftrightarrow C_5H_6N^+ + HCO_3^-$$

A. $C_5H_6N^+$
B. HCO_3^-
C. C_5H_5N
D. H_2CO_3
E. H_3O^+

53. Which of the following terms applies to Cl^- in the reaction below?

$$HCl\ (aq) \rightarrow H^+ + Cl^-$$

A. Weak conjugate base
B. Strong conjugate base
C. Weak conjugate acid
D. Strong conjugate acid
E. Strong conjugate base and weak conjugate acid

54. Which of the following compounds is a diprotic acid?

A. HCl
B. H_3PO_4
C. HNO_3
D. H_2SO_3
E. H_2O

55. Lysine contains two amine groups ($pK_a = 9.0$ and 10.0) and a carboxylic acid group (pK_a = 2.2). In a solution of pH 9.5, which describes the protonation and charge state of lysine?

A. Carboxylic acid is deprotonated and negative; amine (pK_a 9.0) is deprotonated and neutral, whereby the amine ($pK_a = 10.0$) is protonated and positive
B. Carboxylic acid is deprotonated and negative; amine ($pK_a = 9.0$) is protonated and positive, whereby the amine ($pK_a = 10.0$) is deprotonated and neutral
C. Carboxylic acid is deprotonated and neutral; both amines are protonated and positive
D. Carboxylic acid is deprotonated and negative; both amines are deprotonated and neutral
E. Carboxylic acid is deprotonated and neutral; amine ($pK_a = 9.0$) is deprotonated and neutral, whereby the amine ($pK_a = 10.0$) is protonated and positive

56. Which of the following is the chemical species present in acidic solutions?

A. H_2O^+ (*aq*)
B. H_3O^+ (*l*)
C. H_2O (*aq*)
D. ^-OH (*aq*)
E. H_3O^+ (*aq*)

57. Which compound has a value of K_a that is approximately equal to 10^{-5}?

A. $CH_3CH_2CH_2CO_2H$
B. KOH
C. NaBr
D. HNO_3
E. NH_3

58. Relative to a pH of 7, a solution with a pH of 4 has:

A. 30 times less [H^+]
B. 300 times less [H^+]
C. 1,000 times greater [H^+]
D. 300 times greater [H^+]
E. 30 times greater [H^+]

59. What is the pH of this buffer system if the concentration of undissociated weak acid is equal to the concentration of the conjugate base? (Use the K_a of the buffer = 4.6×10^{-4})

A. 1 and 2
B. 3 and 4
C. 5 and 6
D. 7 and 8
E. 9 and 10

60. Which of the following is the ionization constant expression for water?

A. $K_w = [H_2O] / [H^+]\cdot[^-OH]$
B. $K_w = [H+]\cdot[^-OH] / [H_2O]$
C. $K_w = [H^+]\cdot[^-OH]$
D. $K_w = [H_2O]\cdot[H_2O]$
E. None of the above

Notes for active learning

==

Practice Set 4: Questions 61–80

==

61. Which of the following statements about strong or weak acids is true?

A. A weak acid reacts with a strong base
B. A strong acid does not react with a strong base
C. A weak acid readily forms ions when dissolved in water
D. A weak acid and a strong acid at the same concentration are equally corrosive
E. None of the above

62. What is the value of K_w at 25 °C?

A. 1.0
B. 1.0×10^{-7}
C. 1.0×10^{-14}
D. 1.0×10^{7}
E. 1.0×10^{14}

63. Which of the statements below best describes the following reaction?

$$HNO_3\ (aq) + LiOH\ (aq) \rightarrow LiNO_3\ (aq) + H_2O\ (l)$$

A. Nitric acid and lithium hydroxide solutions produce lithium nitrate solution and H_2O
B. Nitric acid and lithium hydroxide solutions produce lithium nitrate and H_2O
C. Nitric acid and lithium hydroxide produce lithium nitrate and H_2O
D. Aqueous solutions of nitric acid and lithium hydroxide produce aqueous lithium nitrate and H_2O
E. An acid plus a base produces H_2O and a salt

64. The Brønsted-Lowry acid and base in the following reaction are, respectively:

$$NH_4^+ + CN^- \rightarrow NH_3 + HCN$$

A. NH_4^+ and ^-CN
B. ^-CN and HCN
C. NH_4^+ and HCN
D. NH_3 and ^-CN
E. NH_3 and NH_4^+

65. Which would NOT be used to make a buffer solution?

A. H_2SO_4
B. H_2CO_3
C. NH_4OH
D. CH_3COOH
E. Tricine

66. Which of the following is a general property of an acidic solution?

A. Turns litmus paper blue
B. Neutralizes acids
C. Tastes bitter
D. Feels slippery
E. None of the above

67. What is the term for a solution that is a good conductor of electricity?

A. Strong electrolyte
B. Weak electrolyte
C. Non-electrolyte
D. Aqueous electrolyte
E. None of the above

68. Which of the following compounds is an acid?

A. HBr **B.** C_2H_6 **C.** KOH **D.** NaF **E.** $NaNH_2$

69. A metal and a salt solution react only if the metal introduced into the solution is:

A. below the replaced metal in the activity series
B. above the replaced metal in the activity series
C. below hydrogen in the activity series
D. above hydrogen in the activity series
E. equal to the replaced metal in the activity series

70. Which of the following is an example of an Arrhenius base?

I. NaOH (*aq*) II. $Al(OH)_3$ (*s*) III. $Ca(OH)_2$ (*aq*)

A. I only
B. II only
C. III only
D. I and II only
E. I and III only

71. Which of the following is NOT a conjugate acid/base pair?

A. S^{2-} / H_2S

B. HSO_4^- / SO_4^{2-}

C. H_2O / ^-OH

D. PH_4^+ / PH_3

E. All are conjugate acid/base pairs

72. If a buffer is made with the pH below the pK_a of the weak acid, the [base] / [acid] ratio is:

A. equal to 0

B. equal to 1

C. greater than 1

D. less than 1

E. undetermined

73. Which of the following acids listed below has the strongest conjugate base?

Monoprotic Acids	K_a
Acid I	1.3×10^{-8}
Acid II	2.9×10^{-9}
Acid III	4.2×10^{-10}
Acid IV	3.8×10^{-8}

A. I

B. II

C. III

D. IV

E. Not enough data to determine

74. Complete neutralization of phosphoric acid with barium hydroxide, when separated and dried, yields $Ba_3(PO_4)_2$ as one of the products. Which term describes phosphoric acid?

A. Monoprotic acid

B. Diprotic acid

C. Hexaprotic acid

D. Tetraprotic acid

E. Triprotic acid

75. Which is the correct net ionic equation for the hydrolysis reaction of Na_2S?

A. $Na^+ + H_2O \rightarrow NaOH + H_2$

B. $Na^+ + 2\ H_2O \rightarrow NaOH + H_2O^+$

C. $S^{2-} + H_2O \rightarrow 2\ HS^- + OH^-$

D. $S^{2-} + 2\ H_2O \rightarrow HS^- + H_3O^+$

E. $S^{2-} + H_2O \rightarrow OH^- + HS^-$

76. Calculate the pH of 0.0765 M HNO_3.

A. 1.1 **B.** 3.9 **C.** 11.7 **D.** 7.9 **E.** 5.6

77. Which of the following compounds is NOT a strong acid?

A. HBr (*aq*)
B. HNO_3
C. CH_3COOH
D. H_2SO_4
E. HCl (*aq*)

78. Which reaction produces $NiCr_2O_7$ as a product?

A. Nickel (II) hydroxide and dichromic acid
B. Nickel (II) hydroxide and chromic acid
C. Nickelic acid and chromium (II) hydroxide
D. Nickel (II) hydroxide and chromate acid
E. Nickel (II) hydroxide and trichromic acid

79. Which of the following statements describes a Brønsted-Lowry base?

A. Donates protons to other substances
B. Accepts protons from other substances
C. Produces hydrogen ions in aqueous solution
D. Produces hydroxide ions in aqueous solution
E. Accepts hydronium ions from other substances

80. When dissolved in water, the Arrhenius acid/bases KOH, H_2SO_4 and HNO_3 are, respectively:

A. base, acid and base
B. base, base and acid
C. base, acid and acid
D. acid, base and base
E. acid, acid and base

Practice Set 5: Questions 81–100

Questions **81–85** are based on the following titration graph:

The unknown acid is completely titrated with NaOH, as shown on the following titration curve:

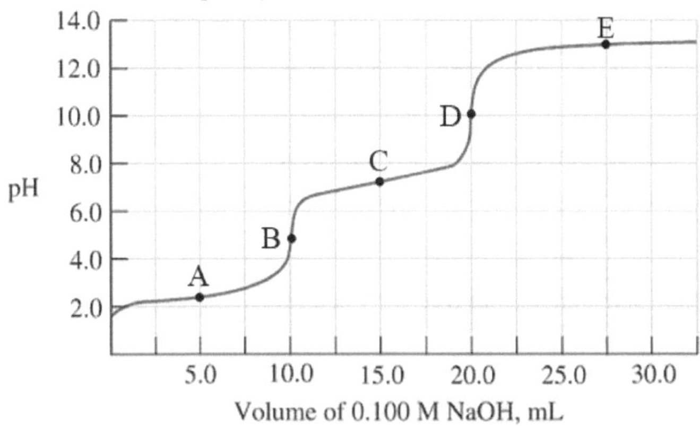

81. The unknown acid shown in the graph must be a(n):

A. monoprotic acid
B. diprotic acid
C. triprotic acid
D. weak acid
E. aprotic acid

82. The pK_{a2} for this acid is located at point:

A. A **B.** B **C.** C **D.** D **E.** E

83. At which point does the acid exist as 50% fully protonated and 50% singly deprotonated?

A. A **B.** B **C.** C **D.** D **E.** E

84. At which point is the acid 100% singly deprotonated?

A. A **B.** B **C.** C **D.** D **E.** E

85. Which points are the best buffer regions?

A. A and B
B. A and C
C. B and D
D. C and B
E. C and D

86. Sodium hydroxide is a strong base. If a concentrated solution of NaOH spills on a latex glove, it feels like water. Why is it that it feels very slippery if the solution were to splash directly on a person's skin?

A. As a liquid, NaOH is slippery, but this cannot be detected through a latex glove because of the friction between the latex surfaces
B. NaOH destroys skin cells on contact, and the remnants of skin cells feel slippery because the cells have been lysed
C. NaOH lifts oil directly out of the skin cells, and the extruded oil causes the slippery sensation
D. NaOH reacts with skin oils, transforming them into soap
E. NaOH, as a liquid, causes the skin to feel slippery from low viscosity

87. Which of the following statements is/are true for a neutralization reaction?

I. Water is formed
II. It is the reaction of an ^-OH with an H^+
III. One molecule of acid neutralizes one molecule of base

A. I only
B. II only
C. III. only
D. I and II only
E. I and III only

88. Which of the following statements is/are true for strong acids?

I. They form positively charged ions when dissolved in H_2O
II. They form negatively charged ions when dissolved in H_2O
III. They are strongly polar

A. I only
B. II only
C. III only
D. II and III only
E. I, II and III

89. A Brønsted-Lowry base is defined as a substance that:

A. increases $[H^+]$ when placed in water
B. decreases $[H^+]$ when placed in water
C. acts as a proton donor
D. acts as a proton acceptor
E. acts as a buffer

90. What is the salt to acid ratio needed to prepare a buffer solution with pH = 4.0 and an acid with a pK_a of 3.0?

A. 1:1
B. 1:1000
C. 1:100
D. 1:5
E. 10:1

91. Which of the following must be true if an unknown solution is a poor conductor of electricity?

A. Solution is slightly reactive
B. Solution is highly corrosive
C. Solution is highly ionized
D. Solution is slightly ionized
E. Solution is highly reactive

92. Which of the following is a general property of a basic solution?

I. Turns litmus paper blue
II. Tastes bitter
III. Feels slippery

A. I only
B. II only
C. III only
D. I and II only
E. I, II and III

93. Since $pK_a = -\log K_a$, which of the following is a correct statement?

A. Since the pK_a for conversion of the ammonium ion to ammonia is 9.3; ammonia is a weaker base than the ammonium ion
B. For carbonic acid with pK_a values of 6.3 and 10.3, the bicarbonate ion is a stronger base than the carbonate ion
C. Acetic acid (pK_a = 4.8) is a weaker acid than lactic acid (pK_a = 3.9)
D. Lactic acid (pK_a = 3.9) is weaker than the forms of phosphoric acid (pK_a = 2.1, 6.9 and 12.4)
E. None of the above

94. Which of the following compounds is NOT a strong acid?

A. HI (*aq*)
B. $HClO_4$
C. HCl (*aq*)
D. HNO_3
E. $HC_2H_3O_2$

95. Which of the following is the acid anhydride for H_3CCOOH?

A. H_3CCOO^-
B. $H_3CCO_4CCH_3$
C. $H_3CCO_3CCH_3$
D. $H_3CCO_2CCH_3$
E. $H_3CCOCCH_3$

96. Acids and bases react to form:

A. Brønsted-Lowry acids
B. Arrhenius acids
C. Lewis acids
D. Lewis bases
E. salts

97. Which of the following does NOT act as a Brønsted-Lowry acid and base?

A. H_2O
B. HCO_3^-
C. $H_2PO_4^-$
D. NH_4^+
E. HS^-

98. Which of the following is the conjugate acid of hydrogen phosphate, HPO_4^{2-}?

A. $H_2PO_4^{2-}$
B. H_3PO_4
C. $H_2PO_4^-$
D. $H_2PO_3^-$
E. None of the above

99. A sample of $Mg(OH)_2$ salt is dissolved in water and reaches equilibrium with its dissociated ions. The addition of the strong base NaOH increases the concentration of:

A. H_2O^+
B. Mg^{2+}
C. undissociated sodium hydroxide
D. undissociated magnesium hydroxide
E. H_2O^+ and undissociated sodium hydroxide

100. Which of the following is the weakest acid?

A. HCO_3^-; $pK_a = 10.3$
B. $HC_2H_3O_2$; $pK_a = 4.8$
C. NH_4^+; $pK_a = 9.2$
D. $H_2PO_4^-$; $pK_a = 7.2$
E. CCl_3COOH; $pK_a = 2.9$

==

Practice Set 6: Questions 101–120

==

101. In which of the following pairs of acids are both chemical species weak acids?

A. $HC_2H_3O_2$ and HI
B. H_2CO_3 and HBr
C. HCN and H_2S
D. H_3PO_4 and H_2SO_4
E. HCl and HBr

102. Which of the following indicators are yellow in an acidic solution and blue in a basic solution?

I. methyl red II. phenolphthalein III. bromothymol blue

A. I only
B. II only
C. I and III only
D. III only
E. I, II and III

103. Which of the following compounds is a basic anhydride?

A. BaO **B.** O_2 **C.** CO_2 **D.** SO_2 **E.** N_2O_5

104. What is the pK_a of an unknown acid if, in a solution at pH 7.0, 24% of the acid is in its deprotonated form?

A. 6.0 **B.** 6.5 **C.** 7.5 **D.** 8.0 **E.** 10.0

105. When acids and bases react, which of the following, other than water, is the product?

A. hydronium ion
B. metal
C. hydrogen ion
D. hydroxide ion
E. salt

106. Which of the following acids listed below is the strongest acid?

Monoprotic Acids	K_a
Acid I	1.0×10^{-8}
Acid II	1.8×10^{-9}
Acid III	3.7×10^{-10}
Acid IV	4.6×10^{-8}

A. Acid I
B. Acid II
C. Acid III
D. Acid IV
E. Requires more information

107. In a basic solution, the pH is [...] and the $[H_3O^+]$ is [...].

A. < 7 and $> 1 \times 10^{-7}$ M
B. < 7 and $< 1 \times 10^{-7}$ M
C. $= 7$ and 1×10^{-7} M
D. < 7 and 1×10^{-7} M
E. > 7 and $< 1 \times 10^{-7}$ M

108. If a battery acid solution is a strong electrolyte, which of the following must be true of the battery acid?

A. It is highly ionized
B. It is slightly ionized
C. It is highly reactive
D. It is slightly reactive
E. It is slightly ionized and weakly reactive

109. Which species form in the second step of the dissociation of H_3PO_4?

A. PO_4^{3-}
B. $H_2PO_4^{2-}$
C. $H_2PO_4^-$
D. HPO_4^{2-}
E. H_3PO_3

110. Which statement concerning the Arrhenius acid–base theory is NOT correct?

A. Neutralization reactions produce H_2O, plus a salt
B. Acid–base reactions must take place in an aqueous solution
C. Arrhenius acids produce H^+ in H_2O solution
D. Arrhenius bases produce ^-OH in H_2O solution
E. All are correct

111. What is the term for a substance that donates a proton in an acid–base reaction?

A. Brønsted-Lowry acid
B. Brønsted-Lowry base
C. Arrhenius acid
D. Arrhenius base
E. Lewis acid

112. Which of the following does NOT represent a conjugate acid/base pair?

A. HCl / Cl^-
B. $HC_2H_3O_2$ / ^-OH
C. H_3O^+ / H_2O
D. HCN / ^-CN
E. All are conjugate acid/base pairs

113. Which of the following statements is correct for a solution of 100% H_2O?

A. It contains no ions
B. It is an electrolyte
C. The [^-OH] equals [H_3O^+]
D. The [^-OH] is greater than [H_3O^+]
E. The [H_3O^+] is greater than [^-OH]

114. The isoelectric point of an amino acid is defined as the pH at which the:

A. value is equal to the pK_a
B. amino acid exists in the acidic form
C. amino acid exists in the basic form
D. amino acid exists in the zwitterion form
E. amino acid exists in the protonated form

115. What is the term for ions that do not participate in a reaction and do not appear in the net ionic equation in an aqueous solution?

A. Zwitterions
B. Spectator ions
C. Nonelectrolyte ions
D. Electrolyte ions
E. None of the above

116. Salts that result from the reaction of strong acids with strong bases are:

A. neutral
B. basic
C. acidic
D. salts
E. none of the above

117. Which of the following pairs of chemical species contains two polyprotic acids?

A. HNO_3 and $H_2C_4H_4O_6$
B. $HC_2H_3O_2$ and $H_3C_6H_5O_7$
C. H_3PO_4 and HCN
D. H_2S and H_2CO_3
E. HCN and HNO_3

118. Which of the following pairs of acids and conjugate bases is NOT correctly labeled?

Acid ↔ Conjugate Base

A. $NH_4^+ \leftrightarrow NH_3$
B. $HSO_3^- \leftrightarrow SO_3^{2-}$
C. $H_2SO_4 \leftrightarrow HSO_4^-$
D. $HSO_4^- \leftrightarrow SO_4^{2-}$
E. $HFO_2 \leftrightarrow HFO_3$

119. What happens to the respective corrosive properties of an acid and base after a neutralization reaction?

A. The corrosive properties are doubled because the acid and base are combined in the salt
B. The corrosive properties remain the same when the salt is mixed into water
C. The corrosive properties are neutralized because the acid and base are transformed
D. The corrosive properties are unaffected because salt is a corrosive agent
E. The corrosive properties are increased because salt is a corrosive agent

120. What is the molarity of a nitric acid solution if 25.00 mL of HNO_3 is required to neutralize 0.500 g of calcium carbonate? (Use molecular mass of $CaCO_3$ = 100.09 g/mol)

$$2\ HNO_3\ (aq) + CaCO_3\ (s) \rightarrow Ca(NO_3)_2\ (aq) + H_2O\ (l) + CO_2\ (g)$$

A. 0.200 M
B. 0.250 M
C. 0.400 M
D. 0.550 M
E. 0.700 M

Notes or active learning

Notes or active learning

Answer Key & Detailed Explanations

Answer Key

1: A	21: E	41: A	61: A	81: B	101: C
2: B	22: A	42: A	62: C	82: C	102: D
3: A	23: D	43: D	63: D	83: A	103: A
4: C	24: D	44: E	64: A	84: B	104: C
5: C	25: B	45: C	65: A	85: B	105: E
6: C	26: B	46: E	66: E	86: D	106: D
7: E	27: B	47: A	67: A	87: E	107: E
8: E	28: D	48: D	68: A	88: E	108: A
9: B	29: C	49: B	69: B	89: D	109: D
10: B	30: C	50: B	70: E	90: E	110: E
11: B	31: E	51: C	71: A	91: D	111: A
12: E	32: C	52: A	72: D	92: E	112: B
13: C	33: A	53: A	73: C	93: C	113: C
14: B	34: D	54: D	74: E	94: E	114: D
15: C	35: E	55: A	75: E	95: C	115: B
16: A	36: C	56: E	76: A	96: E	116: A
17: D	37: B	57: A	77: C	97: D	117: D
18: C	38: B	58: C	78: A	98: C	118: E
19: D	39: B	59: B	79: B	99: D	119: C
20: C	40: E	60: C	80: C	100: A	120: C

==

Practice Set 1: Questions 1–20

==

1. A is correct.

An acid dissociates a proton to form the conjugate base, while a conjugate base accepts a proton to form the acid.

2. B is correct.

pH = –log[H^+]

pH = –log(0.10)

pH = 1

3. A is correct.

Salt is a combination of an acid and a base.

Its pH will be determined by the acid and base that created the salt.

Combinations include:

Weak acid and strong base: salt is basic

Strong acid and weak base: salt is acidic

Strong acid and strong base: salt is neutral

Weak acid and weak base: could be anything (acidic, basic, or neutral)

Examples of strong acids and bases that commonly appear in chemistry problems:

Strong acids: HCl, HBr, HI, H_2SO_4, $HClO_4$ and HNO_3

Strong bases: NaOH, KOH, $Ca(OH)_2$ and $Ba(OH)_2$

The hydroxides of Group I and II metals are considered strong bases.

Examples include LiOH (lithium hydroxide), NaOH (sodium hydroxide), KOH (potassium hydroxide), RbOH (rubidium hydroxide), CsOH (cesium hydroxide), $Ca(OH)_2$ (calcium hydroxide), $Sr(OH)_2$ (strontium hydroxide) and $Ba(OH)_2$ (barium hydroxide).

4. C is correct.

The Brønsted-Lowry acid–base theory focuses on the ability to accept and donate protons (H^+).

A Brønsted-Lowry acid is a term for a substance that donates a proton (H^+) in an acid–base reaction, while a Brønsted-Lowry base is a substance that accepts a proton.

5. C is correct.

A solution's *conductivity* is correlated to the number of ions present in a solution.

The bulb is shining brightly implies that the solution is an excellent conductor; meaning that the solution has a high concentration of ions.

6. C is correct.

An amphoteric compound can react as an acid (i.e., donates protons) as well as a base (i.e., accepts protons).

One type of amphoteric species is amphiprotic molecules, which can either accept or donate a proton (H^+).

Examples of amphiprotic molecules include amino acids (i.e., an amine and carboxylic acid group) and self-ionizable compounds such as water.

7. E is correct.

Strong acids dissociate protons into the aqueous solution. The resulting anion is stable, which accounts for the ~100% ionization of the strong acid.

Weak acids do not completely (or appreciably) dissociate protons into the aqueous solution. The resulting anion is unstable, which accounts for a small ionization of the weak acid.

Each of the listed strong acids forms an anion stabilized by resonance.

$HClO_4$ (perchloric acid) has a pK_a of –10.

H_2SO_4 (sulfuric acid) is a diprotic acid with a pK_a of –3 and 1.99.

HNO_3 (nitrous acid) has a pK_a of –1.4.

8. E is correct.

To find the pH of strong acid and strong base solutions (where $[H^+] > 10^{-6}$), use the equation:

$$pH = -\log[H_3O^+]$$

Given that $[H^+] < 10^{-6}$, the pH = 4.

Note that this approach only applies to strong acids and strong bases.

9. B is correct.

The self-ionization (autoionization) of water is an ionization reaction in pure water or an aqueous solution, in which a water molecule, H_2O, deprotonates (loses the nucleus of one of its hydrogen atoms) to become a hydroxide ion (^-OH).

10. B is correct.

An electrolyte is a substance that dissociates into cations (i.e., positive ions) and anions (i.e., negative ions) when placed in solution.

NH_3 is a weak acid with a pK_a of 38. Therefore, it does not readily dissociate into H^+ and $^-NH_2$.

HCO_3^- (bicarbonate) has a pK_a of about 10.3 and is considered a weak acid because it does not dissociate completely. Therefore, it does not readily dissociate into H^+ and CO_3^{2-}.

HCN (nitrile) is a weak acid with a pK_a of 9.3.

Therefore, it does not readily dissociate into H^+ and ^-CN (i.e., cyanide).

11. B is correct.

If the $[H_3O^+] = [^-OH]$, it has a pH of 7, and the solution is neutral.

12. E is correct.

The Arrhenius acid–base theory states that acids produce H^+ ions (protons) in H_2O solution and bases produce ^-OH ions (hydroxide) in H_2O solution.

13. C is correct.

Brønsted-Lowry acid–base focuses on the ability to accept and donate protons (H^+).

A Brønsted-Lowry acid is a term for a substance that donates a proton in an acid–base reaction, while a Brønsted-Lowry base is a substance that accepts a proton.

The definition is expressed in terms of an equilibrium expression

acid + base ↔ conjugate base + conjugate acid.

14. B is correct.

According to the Brønsted-Lowry acid–base theory:

An acid (reactant) dissociates a proton to become the conjugate base (product).

A base (reactant) gains a proton to become the conjugate acid (product).

The definition is expressed as an equilibrium expression

acid + base ↔ conjugate base + conjugate acid.

15. C is correct.

The pH scale ranges from 1 to 14, with 7 being a neutral pH.

Acidic solutions have a pH below 7, while basic solutions have a pH above 7.

The pH scale is a log scale where 7 has a 50% deprotonated and 50% protonated (neutral) species (1:1 ratio).

At a pH of 6, there are 1 deprotonated : 10 protonated species. The ratio is 1:10.

At a pH of 5, there are 1 deprotonated : 100 protonated species. The ratio is 1:100.

A pH change of 1 unit changes the ratio by 10×.

Lower pH (< 7) results in more protonated species (e.g., cation), while an increase in pH (> 7) results in more deprotonated species (e.g., anion).

16. A is correct.

pI is the symbol for isoelectric point: pH where a protein ion has zero net charges.

To calculate pI of amino acids with 2 pK_a values, take the average of the pK_a's:

$$pI = (pK_{a1} + pK_{a2}) / 2$$

$$pI = (2.2 + 4.2) / 2$$

$$pI = 3.2$$

17. D is correct.

Acidic solutions contain hydronium ions (H_3O^+). These ions are in the aqueous form because they are dissolved in water.

Although chemists often write H^+ (*aq*), referring to a single hydrogen nucleus (a proton), it exists as a hydronium ion (H_3O^+).

18. C is correct.

The equivalence point is when chemically equivalent quantities of acid and base have been mixed. The moles of acid are equivalent to the moles of base.

The endpoint (related to but not the same as the equivalence point) refers to the point at which the indicator changes color in a colorimetric titration. The endpoint can be found by an indicator, such as phenolphthalein (i.e., it turns colorless in acidic solutions and pink in basic solutions).

A buffer is an aqueous solution that consists of a weak acid and its conjugate base, or vice versa. Buffered solutions resist changes in pH and are often used to keep the pH at a nearly constant value in many chemical applications. It does this by readily absorbing or releasing protons (H^+) and ^-OH.

When an acid is added to the solution, the buffer releases ^-OH and accepts H^+ ions from the acid. When a base is added, the buffer accepts ^-OH ions from the base and releases protons (H^+).

Using the Henderson-Hasselbalch equation: $pH = pK_a + \log([A^-] / [HA])$

where [HA] = concentration of the weak acid, in units of molarity; $[A^-]$ = concentration of the conjugate base, in units of molarity

$$pK_a = -\log(K_a)$$

where K_a = acid dissociation constant.

From the question, the concentration of the conjugate base equals the concentration of the weak acid. This equates to the following expression:

$$[HA] = [A^-]$$

Rearranging: $[A^-] / [HA]$, like in the Henderson-Hasselbalch equation:

$$[A^-] / [HA] = 1$$

Substituting the value into the Henderson-Hasselbalch equation:

$$pH = pK_a + \log(1)$$

$$pH = pK_a + 0$$

$$pH = pK_a$$

When $pH = pK_a$, the titration is in the buffering region.

19. D is correct.

The ionic product constant of water:

$$K_w = [H_3O^+]\cdot[^-OH]$$

$$[^-OH] = K_w / [H_3O^+]$$

$$[^-OH] = [1 \times 10^{-14}] / [7.5 \times 10^{-9}]$$

$$[^-OH] = 1.3 \times 10^{-6}$$

20. C is correct.

Consider the K_a presented in the question – which species is more acidic or basic than NH_3?

NH_4^+ is correct because it is NH_3 after absorbing one proton. The concentration of H^+ in water is proportional to K_a. NH_4^+ is the conjugate acid of NH_3.

A: H^+ is acidic.

B: NH_2^- is NH_3 with one less proton. If a base loses a proton, it would be an even stronger base with a higher proton affinity, so it is not weaker than NH_3.

D: Water is neutral.

E: $NaNH_2$ is the neutral species of NH_2^-.

Notes for active learning

==

Practice Set 2: Questions 21–40

==

21. E is correct.

Acid as a reactant produces a conjugate base, while a base produces a conjugate acid.

The conjugate base of a chemical species is that species after H^+ has dissociated.

Therefore, the conjugate base of HSO_4^- is SO_4^{2-}.

The conjugate base of H_3O^+ is H_2O.

22. A is correct.

Start by calculating the moles of $Ca(OH)_2$:

Moles of $Ca(OH)_2$ = molarity $Ca(OH)_2$ × volume of $Ca(OH)_2$

Moles of $Ca(OH)_2$ = 0.1 M × (30 mL × 0.001 L/mL)

Moles of $Ca(OH)_2$ = 0.003 mol

Use the coefficients from the reaction equation to determine moles of HNO_3:

Moles of HNO_3 = (coefficient of HNO_3) / [coefficient $Ca(OH)_2$ × moles of $Ca(OH)_2$]

Moles of HNO_3 = (2 / 1) × 0.003 mol

Moles of HNO_3 = 0.006 mol

Divide moles by molarity to calculate volume:

Volume of HNO_3 = moles of HNO_3 / molarity of HNO_3

Volume of HNO_3 = 0.006 mol / 0.2 M

Volume of HNO_3 = 0.03 L

Convert volume to milliliters:

0.03 L × 1000 mL / L = 30 mL

23. D is correct.

Acidic solutions have a pH less than 7 due to a higher concentration of the H^+ ions relative to ^-OH ions.

Basic solutions have a pH greater than 7 due to a higher concentration of ^-OH ions relative to H^+ ions.

24. D is correct.

An amphoteric compound can react as an acid (i.e., donates protons) as well as a base (i.e., accepts protons).

Examples of amphoteric molecules include amino acids (i.e., an amine and carboxylic acid group) and self-ionizable compounds such as water.

25. B is correct.

Strong acids (i.e., reactants) proceed towards products.

$$K_a = [\text{products}] / [\text{reactants}]$$

The molecule with the largest K_a is the strongest acid.

$$pK_a = -\log K_a$$

The molecule with the smallest pK_a is the strongest acid.

Strong acids dissociate a proton to produce the weakest conjugate base (i.e., most stable anion).

Weak acids dissociate a proton to produce the strongest conjugate base (i.e., least stable anion).

26. B is correct.

Balanced reaction:

$$H_3PO_4 + 3\ LIOH = Li_3PO_4 + 3\ H_2O$$

In the neutralization of acids and bases, salt and water result.

Phosphoric acid and lithium hydroxide react, so the resulting compounds are lithium phosphate and water.

27. B is correct.

The stability of the compound determines base strength. If the compound is unstable in its present state, it seeks a bonding partner (e.g., H+ or another atom) by donating its electrons for the new bond formation.

The 8 strong bases are: LiOH (lithium hydroxide), NaOH (sodium hydroxide), KOH (potassium hydroxide), $Ca(OH)_2$ (calcium hydroxide), RbOH (rubidium hydroxide), $Sr(OH)_2$, (strontium hydroxide), CsOH (cesium hydroxide) and $Ba(OH)_2$ (barium hydroxide).

28. D is correct.

The formula for pH:

$$pH = -\log[H^+]$$

Rearrange to solve for $[H^+]$:

$$[H^+] = 10^{-pH}$$

$$[H^+] = 10^{-2}\ M = 0.01\ M$$

29. C is correct.

Ionization	Dissociation
The process that produces new charged particles.	The separation of charged particles that already exist in a compound.
Involves polar covalent compounds or metals.	Involves ionic compounds.
Involves covalent bonds between atoms	Involves ionic bonds in compounds
Produces charged particles.	Produces either charged particles or electrically neutral particles.
Irreversible	Reversible
Example: $HCl \rightarrow H^+ + Cl^-$ $Mg \rightarrow Mg^{2+} + 2\ e^-$	Example: $PbBr_2 \rightarrow Pb^{2+} + 2\ Br^-$

30. C is correct.

A base is a chemical substance with a pH greater than 7 and feels slippery because it dissolves the fatty acids and oils from the skin and reduces the friction between skin cells.

Under acidic conditions, litmus paper is red, and under basic conditions, it is blue.

Many bitter-tasting foods are alkaline because bitter compounds often contain amine groups, which are weak bases.

Acids are known to have a sour taste (e.g., lemon juice) because the sour taste receptors on the tongue detect the dissolved hydrogen (H^+) ions.

31. E is correct.

The 7 strong acids are:

HCl (hydrochloric acid), HNO_3 (nitric acid), H_2SO_4 (sulfuric acid), HBr (hydrobromic acid), HI (hydroiodic acid), $HClO_3$ (chloric acid), and $HClO_4$ (perchloric acid).

The 8 strong bases are:

LiOH (lithium hydroxide), NaOH (sodium hydroxide), KOH (potassium hydroxide), $Ca(OH)_2$ (calcium hydroxide), RbOH (rubidium hydroxide), $Sr(OH)_2$, (strontium hydroxide), CsOH (cesium hydroxide) and $Ba(OH)_2$ (barium hydroxide).

32. C is correct.

The Arrhenius acid–base theory states that acids produce H^+ ions in H_2O solution and bases produce ^-OH ions in H_2O solution.

The Brønsted-Lowry acid–base theory focuses on the ability to accept and donate protons (H^+).

A Brønsted-Lowry acid is a term for a substance that donates a proton in an acid–base reaction, while a Brønsted-Lowry base is a substance that accepts a proton.

Lewis acids are electron-pair acceptors, whereas Lewis bases are electron-pair donors.

33. A is correct.

By the Brønsted-Lowry acid–base theory: an acid (reactant) dissociates a proton to become the conjugate base (product), a base (reactant) gains a proton to become the conjugate acid (product).

The definition is expressed as an equilibrium expression: acid + base ↔ conjugate base + conjugate acid.

34. D is correct.

With polyprotic acids (i.e., more than one H^+ present), the pK_a indicates the pH at which the H^+ is deprotonated. If the pH goes above the first pK_a, one proton dissociates, and so on.

In this example, the pH is above the first and second pK_a, so two acid groups are deprotonated while the third acidic proton is unaffected.

35. E is correct.

It is important to identify the acid that is active in the reaction.

The parent acid is defined as the most protonated form of the buffer. The number of dissociating protons an acid can donate depends on the charge of its conjugate base.

$Ba_2P_2O_7$ is given as one of the products in the reaction. Because barium is a group 2B metal, it has a stable oxidation state of +2. Because two barium cations are present in the product, the charge of P_2O_7 ion (the conjugate base in the reaction) must be –4.

Therefore, the fully protonated form of this conjugate must be $H_4P_2O_7$, which is a tetraprotic acid because it has 4 protons that can dissociate.

36. C is correct.

The greater the concentration of H_3O^+, the more acidic the solution is.

37. B is correct.

A triprotic acid has three protons that can dissociate.

38. B is correct.

KCl and NaI are salts. Two salts react if one of the products precipitates.

In this example, the products (KI and NaCl) are soluble in water, so they do not react.

39. B is correct.

Sodium acetate is a basic compound because acetate is the conjugate base of acetic acid, a weak acid ("the conjugate base of a weak acid acts as a base in the water").

The addition of a base to any solution, even a buffered solution, increases the pH.

40. E is correct.

An acid anhydride is a compound that has two acyl groups bonded to the same oxygen atom.

Anhydride means *without water* and is formed via dehydration (i.e., removal of H_2O).

Notes for active learning

===

Practice Set 3: Questions 41–60

===

41. A is correct.

The Brønsted-Lowry acid–base theory focuses on the ability to accept and donate protons (H^+).

A Brønsted-Lowry acid is a term for a substance that donates a proton (H^+) in an acid–base reaction, while a Brønsted-Lowry base is a substance that accepts a proton.

42. A is correct.

Strong acids (i.e., reactants) proceed towards products.

K_a = [products] / [reactants]

The molecule with the largest K_a is the strongest acid.

$pK_a = -\log K_a$

The molecule with the smallest pK_a is the strongest acid.

Strong acids dissociate a proton to produce the weakest conjugate base (i.e., most stable anion).

Weak acids dissociate a proton to produce the strongest conjugate base (i.e., least stable anion).

43. D is correct.

Learn the ions involved in boiler scale formations: CO_3^{2-} and the metal ions.

44. E is correct.

A buffer is an aqueous solution that consists of a weak acid and its conjugate base, or vice versa.

Buffered solutions resist changes in pH and are often used to keep the pH at a nearly constant value in many chemical applications. It does this by readily absorbing or releasing protons (H^+) and ^-OH.

When an acid is added to the solution, the buffer releases ^-OH and accepts H^+ ions from the acid.

To create a buffer solution, there needs to be a pair of a weak acid/base and its conjugate, or a salt that contains an ion from the weak acid/base.

Since the problem indicates that sulfoxylic acid (H_2SO_2), which has a pK_a 7.97, needs to be in the mixture, the other component would be an HSO_2^- ion (bisulfoxylate) of $NaHSO_2$.

45. C is correct.

Acidic salt is a salt that still contains H^+ in its anion. It is formed when a polyprotic acid is partially neutralized, leaving at least 1 H^+.

For example:

$H_3PO_4 + 2\ KOH \rightarrow K_2HPO_4 + 2\ H_2O$: (partial neutralization, K_2HPO_4 is acidic salt)

While:

$H_3PO_4 + 3\ KOH \rightarrow K_3PO_4 + 3\ H_2O$: (complete neutralization, K_3PO_4 is not acidic salt)

46. E is correct.

An acid anhydride is a compound that has two acyl groups bonded to the same oxygen atom.

Anhydride means *without water* and is formed via dehydration (i.e., removal of H_2O).

47. A is correct.

By the Brønsted-Lowry definition, an acid donates protons, while a base accepts protons.

On the product side of the reaction, H_2O acts as a base (i.e., the conjugate base of H_3O^+), and HCl acts as an acid (i.e., the conjugate acid of Cl^-).

48. D is correct.

The ratio of the conjugate base to the acid must be determined from the pH of the solution and the pK_a of the acidic component in the reaction.

In the reaction, $H_2PO_4^-$ acts as the acid, and HPO_4^{2-} acts as the base, so the pK_a of $H_2PO_4^-$ should be used in the equation.

Substitute the given values into the Henderson-Hasselbalch equation:

$pH = pK_a + \log[salt / acid]$

$7.35 = 6.87 + \log[salt / acid]$

Since $H_2PO_4^-$ is acting as the acid, subtract 6.87 from both sides:

$0.48 = \log[salt / acid]$

The log base is 10, so the inverse log gives:

$10^{0.48} = (salt / acid)$

$(salt / acid) = 3.02$

The conjugate base (or salt) and the acid ratio is 3.02 / 1.

49. B is correct.

An electrolyte is a substance that dissociates into cations (i.e., positive ions) and anions (i.e., negative ions) when placed in solution. The light bulb is dimly lit, indicating that the solution contains only a low concentration (i.e., partial ionization) of the ions.

An electrolyte produces an electrically conducting solution when dissolved in a polar solvent (e.g., water). The dissolved ions disperse uniformly through the solvent.

If an electrical potential (i.e., voltage) is applied to such a solution, the cations of the solution migrate towards the electrode (i.e., an abundance of electrons).

In contrast, anions migrate towards the electrode (i.e., a deficit of electrons).

50. B is correct.

An acid is a chemical substance with a pH less than 7, producing H^+ ions in water. An acid can be neutralized by a base (i.e., a substance with a pH above 7) to form a salt.

Acids are known to have a sour taste (e.g., lemon juice) because the sour taste receptors on the tongue detect the dissolved hydrogen (H^+) ions.

However, acids are not known to have a slippery feel; this is characteristic of bases.

Bases feel slippery because they dissolve the fatty acids and oils from the skin and reduce the friction between the skin cells.

51. C is correct.

The Arrhenius acid–base theory states that acids produce H^+ ions (protons) in H_2O solution and bases produce ^-OH ions (hydroxide) in H_2O solution.

The Brønsted-Lowry acid–base theory focuses on the ability to accept and donate protons (H^+).

A Brønsted-Lowry acid is a term for a substance that donates a proton in an acid–base reaction, while a Brønsted-Lowry base is a substance that accepts a proton.

52. A is correct.

By the Brønsted-Lowry acid–base theory:

An acid (reactant) dissociates a proton to become the conjugate base (product).

A base (reactant) gains a proton to become the conjugate acid (product).

The definition is expressed in terms of an equilibrium expression:

acid + base ↔ conjugate base + conjugate acid.

53. A is correct.

By the Brønsted-Lowry acid–base theory:

An acid (reactant) dissociates a proton to become the conjugate base (product).

A base (reactant) gains a proton to become the conjugate acid (product).

The definition is expressed in terms of an equilibrium expression:

acid + base ↔ conjugate base + conjugate acid.

HCl dissociates completely and therefore is a strong acid.

Strong acids produce weak (i.e., stable) conjugate bases.

54. D is correct.

A diprotic acid has two protons that can dissociate.

55. A is correct.

Protons (H^+) migrate between amino acid and solvent, depending on the pH of solvent and pK_a of functional groups on the amino acid.

Carboxylic acid groups can donate protons, while amine groups can receive protons.

For the carboxylic acid group:

If the pH of solution < pK_a : group is protonated and neutral

If the pH of solution > pK_a : group is deprotonated and negative

For the amine group:

If the pH of solution < pK_a : group is protonated and positive

If the pH of solution > pK_a : group is deprotonated and neutral

56. E is correct.

Acidic solutions contain hydronium ions (H_3O^+). These ions are in the aqueous form because they are dissolved in water. Although chemists often write H^+ (*aq*), referring to a single hydrogen nucleus (a proton), it exists as the hydronium ion (H_3O^+).

57. A is correct.

With a K_a of 10^{-5}, the pH of a 1 M solution of the carboxylic acid, CH_3CH_2CH2CO_2H, would be 5 and is a weak acid.

$CH_3CH_2CH_2CO_2H$ can be considered a weak acid because it yields a (relatively) unstable anion.

58. C is correct.

Condition 1: pH = 4

$[H_3O^+] = 10^{-pH} = 10^{-4}$

Condition 2: pH = 7

$[H_3O^+] = 10^{-pH} = 10^{-7}$

Ratio of $[H_3O^+]$ in condition 1 and condition 2:

$10^{-4} : 10^{-7} = 1{,}000 : 1$

Solution with a pH of 4 has 1,000 times greater $[H^+]$ than a solution with a pH of 7.

Note: $[H_3O^+]$ is equivalent to $[H^+]$

59. B is correct.

The pH of a buffer is calculated using the Henderson-Hasselbalch equation:

$pH = pK_a + \log([\text{conjugate base}] / [\text{conjugate acid}])$

When [acid] = [base], the fraction is 1.

Log 1 = 0,

$pH = pK_a + 0$

If the K_a of the acid is 4.6×10^{-4} (between 10^{-4} and 10^{-3}), the pK_a (and therefore the pH) is between 3 and 4.

60. C is correct.

$K_w = [H^+]\cdot[^-OH]$ is the definition of the ionization constant for water.

Notes for active learning

==

Practice Set 4: Questions 61–80

==

61. A is correct.

Weak acids react with a strong base, converted to its weak conjugate base, creating an overall basic solution.

Strong acids do react with a strong base, creating a neutral solution.

Weak acids partially dissociate when dissolved in water; unlike a strong acid, they do not readily form ions.

Strong acids are much more corrosive than weak acids.

62. C is correct.

K_w is the water ionization constant (or *water autoprotolysis constant*).

It can be determined experimentally and equals 1.011×10^{-14} at 25 °C (1.00×10^{-14} is used).

63. D is correct.

All options are correct descriptions of the reaction, but the correct choice is the most descriptive.

64. A is correct.

The Brønsted-Lowry acid–base theory focuses on the ability to accept and donate protons (H^+).

A Brønsted-Lowry acid is a term for a substance that donates a proton in an acid–base reaction, while a Brønsted-Lowry base is a substance that accepts a proton.

65. A is correct.

Buffered solutions resist changes in pH and are often used to keep the pH at a nearly constant value in many chemical applications. It does this by readily absorbing or releasing protons (H^+) and ^-OH.

H_2SO_4 is a strong acid. Weak acids and their salts are good buffers.

A buffer is an aqueous solution that consists of a weak acid and its conjugate base, or vice versa.

When an acid is added to the solution, the buffer releases ^-OH and accepts H^+ ions from the acid.

When a base is added, the buffer accepts ^-OH ions from the base and releases protons (H^+).

66. E is correct.

The main use of litmus paper is to test whether a solution is acidic or basic.

Litmus paper can be used to test for water-soluble gases that affect acidity or alkalinity; the gas dissolves in the water, and the resulting solution colors the litmus paper. For example, alkaline ammonia gas causes the litmus paper to change from red to blue.

Blue litmus paper turns red under acidic conditions, and red litmus paper turns blue under basic or alkaline conditions, with the color change occurring over the pH range 4.5-8.3 at 25 °C (77 °F).

Neutral litmus paper is purple.

Litmus can be prepared as an aqueous solution that functions similarly. Under acidic conditions, the solution is red, and under basic conditions, the solution is blue.

The properties listed above (turning litmus paper blue, bitter taste, slippery feel, and neutralizing acids) are true of bases.

An acidic solution has opposite qualities.

It has a pH lower than 7 and therefore turns litmus paper red.

It neutralizes bases, tastes sour, and does not feel slippery.

67. A is correct.

An electrolyte is a substance that dissociates into cations (i.e., positive ions) and anions (i.e., negative ions) when placed in solution.

An electrolyte produces an electrically conducting solution when dissolved in a polar solvent (e.g., water). The dissolved ions disperse uniformly through the solvent.

If an electrical potential (i.e., voltage) is applied to such a solution, the cations of the solution migrate towards the electrode (i.e., an abundance of electrons).

In contrast, the anions migrate towards the electrode (i.e., a deficit of electrons).

An acid is a substance that ionizes when dissolved in suitable ionizing solvents such as water.

If a high proportion of the solute dissociates to form free ions, it is a strong electrolyte.

If most of the solute does not dissociate, it is a weak electrolyte.

The more free ions present, the better the solution conducts electricity.

68. A is correct.

The 7 strong acids are HCl (hydrochloric acid), HNO_3 (nitric acid), H_2SO_4 (sulfuric acid), HBr (hydrobromic acid), HI (hydroiodic acid), $HClO_3$ (chloric acid), and $HClO_4$ (perchloric acid).

69. B is correct.

The activity series determines if a metal displaces another metal in the solution.

The reaction can only occur if the added metal is above (i.e., activity series) the metal currently bonded with the anion. An activity series ranks substances in their order of relative reactivity.

For example, magnesium metal can displace hydrogen ions from solution, so it is more reactive than elemental hydrogen:

$Mg\ (s) + 2\ H^+\ (aq) \rightarrow H_2\ (g) + Mg^{2+}\ (aq)$

Zinc can displace hydrogen ions from solution, so zinc is more reactive than elemental hydrogen:

$Zn\ (s) + 2\ H^+\ (aq) \rightarrow H_2\ (g) + Zn^{2+}\ (aq)$

Magnesium metal can displace zinc ions from solution:

$Mg\ (s) + Zn^{2+}\ (aq) \rightarrow Zn\ (s) + Mg^{2+}\ (aq)$

The metal activity series with the most active (i.e., most strongly reducing) metals appear at the top, and the least active metals near the bottom.

Li: $2\ Li\ (s) + 2\ H_2O\ (l) \rightarrow LiOH\ (aq) + H_2\ (g)$

K: $2\ K\ (s) + 2\ H_2O\ (l) \rightarrow 2\ KOH\ (aq) + H_2\ (g)$

Ca: $Ca\ (s) + 2\ H_2O\ (l) \rightarrow Ca(OH)_2\ (s) + H_2\ (g)$

Na: $2\ Na\ (s) + 2\ H_2O\ (l) \rightarrow 2\ NaOH\ (aq) + H_2\ (g)$

The above can displace H_2 from water, steam, or acids

Mg: $Mg\ (s) + 2\ H_2O\ (g) \rightarrow Mg(OH)_2\ (s) + H_2\ (g)$

Al: $2\ Al\ (s) + 6\ H_2O\ (g) \rightarrow 2\ Al(OH)_3\ (s) + 3\ H_2\ (g)$

Mn: $Mn\ (s) + 2\ H_2O\ (g) \rightarrow Mn(OH)_2\ (s) + H_2\ (g)$

Zn: $Zn\ (s) + 2\ H_2O\ (g) \rightarrow Zn(OH)_2\ (s) + H_2\ (g)$

Fe: $Fe\ (s) + 2\ H_2O\ (g) \rightarrow Fe(OH)_2\ (s) + H_2\ (g)$

The above can displace H_2 from steam or acids

Ni: $Ni\ (s) + 2\ H^+\ (aq) \rightarrow Ni^{2+}\ (aq) + H_2\ (g)$

Sn: $Sn\ (s) + 2\ H^+\ (aq) \rightarrow Sn^{2+}\ (aq) + H_2\ (g)$

Pb: $Pb\ (s) + 2\ H^+\ (aq) \rightarrow Pb^{2+}\ (aq) + H_2\ (g)$

The above can displace H_2 from acids only

$H_2 > Cu > Ag > Pt > Au$

The above cannot displace H_2

70. E is correct.

The Arrhenius acid–base theory states that acids produce H^+ ions (protons) in H_2O solution, and bases produce ^-OH ions (hydroxide) in H_2O solution.

$Al(OH)_3\ (s)$ is insoluble in water; therefore, it cannot function as an Arrhenius base.

71. A is correct.

By the Brønsted-Lowry acid–base theory:

An acid (reactant) dissociates a proton to become the conjugate base (product).

A base (reactant) gains a proton to become the conjugate acid (product).

The definition is expressed in terms of an equilibrium expression:

acid + base ↔ conjugate base + conjugate acid.

72. D is correct.

Henderson-Hasselbach equation:

$pH = pK_a + \log(A^- / HA)$

A buffer is an aqueous solution that consists of a weak acid and its conjugate base, or vice versa.

Buffered solutions resist changes in pH and are often used to keep the pH at a nearly constant value in many chemical applications. It does this by readily absorbing or releasing protons (H^+) and ^-OH.

When an acid is added to the solution, the buffer releases ^-OH and accepts H^+ ions from the acid.

73. C is correct.

The weakest acid has the smallest K_a (or largest pK_a).

The weakest acid has the strongest (i.e., least stable) conjugate base.

74. E is correct.

The balanced reaction:

$$2\ H_3PO_4 + 3\ Ba(OH)_2 \rightarrow Ba_3(PO_4)_2 + 6\ H_2O$$

There are 2 moles of H_3PO_4 in a balanced reaction.

However, acids are categorized by the number of H^+ per mole of acid.

For example:

HCl is monoprotic (has one H^+ to dissociate).

H_2SO_4 is diprotic (has two H^+ to dissociate).

H_3PO_4 is a triprotic acid (has three H^+ to dissociate).

75. E is correct.

Because Na^+ forms a strong base (NaOH) and S forms a weak acid (H_2S), it undergoes hydrolysis in water:

$$Na_2S\ (aq) + 2\ H_2O\ (l) \rightarrow 2\ NaOH\ (aq) + H_2S$$

Ionic equation for individual ions:

$$2\ Na^+ + S^{2-} + 2\ H_2O \rightarrow 2\ Na^+ + 2\ OH^- + H_2S$$

Removing Na^+ ions from both sides of the reaction:

$$S^{2-} + 2\ H_2O \rightarrow 2\ OH^- + H_2S$$

Removing H_2O from both sides of the reaction:

$$S^{2-} + H_2O \rightarrow OH^- + HS^-$$

76. A is correct.

HNO_3 is a strong acid, which means that $[HNO_3] = [H_3O^+] = 0.0765$.

$$pH = -\log[H_3O^+]$$

$$pH = -\log(0.0765)$$

$$pH = 1.1$$

77. C is correct.

Strong acids dissociate a proton to produce the weakest conjugate base (i.e., most stable anion).

Weak acids dissociate a proton to produce the strongest conjugate base (i.e., least stable anion).

Acetic acid (CH_3COOH) has a pK_a of about 4.8 and is considered a weak acid because it does not dissociate completely. Therefore, it does not readily dissociate into H^+ and CH_3COO^-.

Each of the listed strong acids forms an anion stabilized by resonance.

HBr (hydrobromic acid) has a pK_a of –9.

HNO_3 (nitrous acid) has a pK_a of –1.4.

H_2SO_4 (sulfuric acid) is a diprotic acid with a pK_a of –3 and 1.99.

HCl (hydrochloric acid) has a pK_a of –6.3.

78. A is correct.

It is important to recognize the chromate ions:

Dichromate: $Cr_2O_7^{2-}$

Chromium (II): Cr^{2+}

Chromic/Chromate: CrO_4^-

Trichromic acid: does not exist.

79. B is correct.

Brønsted-Lowry acid–base theory focuses on the ability to accept and donate protons (H^+).

A Brønsted-Lowry acid is a term for a substance that donates a proton in an acid–base reaction, while a Brønsted-Lowry base is a substance that accepts a proton.

80. C is correct.

In the Arrhenius theory, acids dissociate in an aqueous solution to produce H^+ (hydrogen ions).

In the Arrhenius theory, bases dissociate in an aqueous solution to produce ^-OH (hydroxide ions).

==

Practice Set 5: Questions 81–100

==

81. B is correct.

The acid requires two equivalents of base to be fully titrated and therefore is a diprotic acid (see diagram).

Using the fully protonated sulfuric acid (H_2SO_4) as an example:

At point A, the acid is 50% fully protonated, 50% singly deprotonated: (50% H_2SO_4: 50% HSO_4^-).

At point B, the acid exists in the singly deprotonated form only (100% HSO_4^-).

At point C, the acid exists as 50% singly deprotonated HSO_4^- and 50% doubly deprotonated SO_4^{2-}.

At point D, the acid exists as 100% SO_4^{2-}.

Point A is pK_{a1}, and point C is pK_{a2} (i.e., strongest buffering regions).

Point B and D are known as equivalence points (i.e., weakest buffering region).

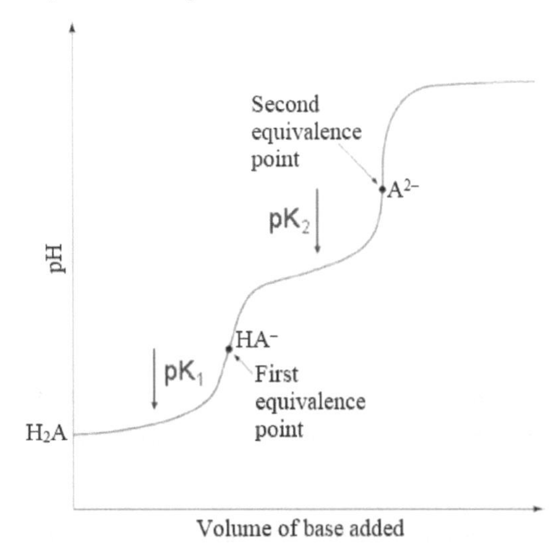

Titration curve: addition of a strong base to diprotic acid (H_2A)

82. C is correct.

Point C on the graph for the question is pK_{a2}. At this point, the acid exists as 50% singly deprotonated (e.g., HSO_4^-) and 50% doubly deprotonated (e.g., SO_4^{2-}).

The Henderson-Hasselbalch equation:

$$pH = pK_{a2} + \log[\text{salt} / \text{acid}]$$

$$pH = pK_{a2} + \log[50\% / 50\%]$$

$$pH = pK_{a2} + \log[1]$$

$$pH = pK_{a2} + 0$$

$$pH = pK_{a2}$$

83. A is correct.

Point A is pK_{a1}. At this point, the acid exists as 50% fully protonated (e.g., H_2SO_4) and 50% singly deprotonated (e.g., HSO_4^-).

The Henderson-Hasselbalch equation:

$$pH = pK_{a1} + \log[salt / acid]$$

$$pH = pK_{a1} + \log[50\% / 50\%]$$

$$pH = pK_{a1} + \log[1]$$

$$pH = pK_{a1} + 0$$

$$pH = pK_{a1}$$

84. B is correct.

At point B on the graph (i.e., equivalence point), the acid exists in the singly deprotonated form only (100% HSO_4^- for the example of H_2SO_4).

85. B is correct.

The buffer region is the flattest region on the curve (the region that resists pH increases with added base). This diprotic acid has two buffering regions: pK_{a1} (around point A) and pK_{a2} (around point C).

A buffer is an aqueous solution that consists of a weak acid and its conjugate base, or vice versa.

Buffered solutions resist changes in pH and are often used to keep the pH at a nearly constant value in many chemical applications. It does this by readily absorbing or releasing protons (H^+) and ^-OH.

When an acid is added to the solution, the buffer releases ^-OH and accepts H^+ ions from the acid.

When a base is added, the buffer accepts ^-OH ions from the base and releases protons (H^+).

86. D is correct.

Sodium hydroxide turns into soap (i.e., saponification) from the reaction with the fatty acid esters and oils on the fingertips (i.e., skin).

Fatty acid esters react with NaOH by releasing free fatty acids, which act as soap surrounding grease with their nonpolar (i.e., hydrophobic) ends, while their polar (i.e., hydrophilic) ends orient towards water molecules. This decreases friction and accounts for the slippery feel of NaOH interacting with the skin.

87. E is correct.

I: water is produced in a neutralization reaction.

II: this reaction is a common example of neutralization, but it is not always the case. For example, weak bases react with strong acids: $AH + B \rightleftharpoons A^- + BH^+$

Or strong bases react with weak acids: $AH + H_2O \rightleftharpoons H_3O^+ + A^-$

III: neutralization occurs when acid donates a proton to the base.

88. E is correct.

Strong acids dissociate entirely, or almost completely, when in water. They dissociate into a positively-charged hydrogen ion (H^+) and another negatively charged ion.

An example is hydrochloric acid (HCl), dissociating into H^+ and Cl^- ions.

Polarity refers to the distribution of electrons in a bond. If a molecule is polar, one side has a partial positive charge, and the other has a partial negative charge.

The more polar the bond, the easier it is for a molecule to dissociate into ions, and therefore the acid is more strongly acidic.

89. D is correct.

The Brønsted-Lowry acid–base theory focuses on the ability to accept and donate protons (H^+).

A Brønsted-Lowry acid is a term for a substance that donates a proton in an acid–base reaction,
while a Brønsted-Lowry base is a substance that accepts a proton.

90. E is correct.

Formula to calculate pH of an acidic buffer:

$$[H^+] = ([\text{acid}] / [\text{salt}]) \times K_a$$

Calculate H^+ from pH:

$$[H^+] = 10^{-pH}$$

$$[H^+] = 10^{-4}$$

Calculate Ka from pK_a:

$$K_a = 10^{-pKa}$$

$$K_a = 10^{-3}$$

Substitute values into the buffer equation:

$[H^+] = ([acid] / [salt]) \times K_a$

$10^{-4} = ([acid] / [salt]) \times 10^{-3}$

$10^{-1} = [acid] / [salt]$

$[salt] / [acid] = 10$

Therefore, the ratio of salt to acid is 10:1.

91. D is correct.

An electrolyte is a substance that dissociates into cations (i.e., positive ions) and anions (i.e., negative ions) when placed in solution. An electrolyte produces an electrically conducting solution when dissolved in a polar solvent (e.g., water). The dissolved ions disperse uniformly through the solvent.

If an electrical potential (i.e., voltage) is applied to such a solution, the cations of the solution migrate towards the electrode (i.e., an abundance of electrons).

In contrast, the anions migrate towards the electrode (i.e., a deficit of electrons).

If a high proportion of the solute dissociates to form free ions, it is a strong electrolyte.

If most of the solute does not dissociate, it is a weak electrolyte.

The more free ions present, the better the solution conducts electricity.

92. E is correct.

Litmus (i.e., dyes extracted from lichens) tests whether a solution is acidic or basic.

Wet litmus paper can be used to test for water-soluble acidic or alkaline gases that dissolve in the water and produce color changes in the litmus paper.

Neutral litmus paper is purple, with color changes occurring over the pH range of 4.5 to 8.3.

Red litmus paper turns blue when exposed to alkaline ammonia gas (basic conditions).

Blue litmus paper turns red under acidic conditions.

93. C is correct.

The more acidic molecule has a lower pK_a.

94. E is correct.

Acetic acid ($CH_3COOH = HC_2H_3O_2$) is commonly known as vinegar. It has a pK_a of about 4.8 and is considered a weak acid because it does not dissociate completely.

HI (hydroiodic acid) has a pK_a of –10.

$HClO_4$ (perchloric acid) has a pK_a of –10.

HCl (hydrochloric acid) has a pK_a of –6.3.

HNO_3 (nitrous acid) has a pK_a of –1.4.

95. C is correct.

An acid anhydride is a compound that has two acyl groups bonded to the same oxygen atom.

$H_3CCO_3CCH_3$

Anhydride means *without water* and is formed via dehydration (i.e., removal of H_2O).

96. E is correct.

Acids and bases react to form H_2O and salts: $NaOH + HCl \rightleftharpoons Na^+Cl^-$ (a salt) + H_2O (water)

The Brønsted-Lowry acid–base theory focuses on the ability to accept and donate protons (H^+).

A Brønsted-Lowry acid is a substance that donates a proton in an acid–base reaction, while a Brønsted-Lowry base is a substance that accepts a proton.

The Arrhenius acid–base theory states that acids produce H^+ ions in H_2O solution and bases produce ^-OH ions in H_2O solution.

Lewis acids are electron-pair acceptors, whereas Lewis bases are electron-pair donors.

97. D is correct.

The Bronsted-Lowry acid–base theory focuses on the ability to accept and donate protons.

The definition is expressed in terms of an equilibrium expression:

acid + base $\leftrightarrow$ conjugate base + conjugate acid.

With an acid, HA, the equation can be written symbolically as:

$HA + B \leftrightarrow A^- + HB^+$

98. C is correct.

By the Brønsted-Lowry acid–base theory:

An acid (reactant) dissociates a proton to become the conjugate base (product).

A base (reactant) gains a proton to become the conjugate acid (product).

The definition is expressed in terms of an equilibrium expression:

acid + base ↔ conjugate base + conjugate acid.

99. D is correct.

The addition of hydroxide (i.e., NaOH) decreases the solubility of magnesium hydroxide due to the common ion effect. Therefore, the amount of undissociated $Mg(OH)_2$ increases.

100. A is correct.

The weakest acid has the highest pK_a (and lowest K_a).

==

Practice Set 6: Questions 101–120

==

101. C is correct.

Strong acids dissociate completely (or almost completely) because they form a stable anion.

Hydrogen cyanide (HCN) has a pK_a of 9.3, and hydrogen sulfide (H_2S) has a pK_a of 7.0.

In this example, the strong acids include HI, HBr, HCl, H_3PO_4 and H_2SO_4.

102. D is correct.

Bromothymol blue is a pH indicator often used for solutions with neutral pH near 7 (e.g., managing the pH of pools and fish tanks). Bromothymol blue acts as a weak acid in a solution that can be protonated or deprotonated. It appears yellow when protonated (lower pH), blue when deprotonated (higher pH), and bluish-green in neutral solution.

Methyl red has a pK_a of 5.1 and is a pH indicator dye that changes color in acidic solutions: it turns red in pH under 4.4, orange in pH between 4.4 and 6.2, and yellow in pH over 6.2.

Phenolphthalein is used as an indicator for acid–base titrations. It is a weak acid, dissociating protons (H^+ ions) in solution. The phenolphthalein molecule is colorless, and the phenolphthalein ion is pink. It turns colorless in acidic solutions and pink in basic solutions.

With basic conditions, the phenolphthalein (neutral) ⇌ ions (pink) equilibrium shifts to the right, leading to more ionization as H^+ ions are removed.

103. A is correct.

Anhydride means *without water* and is formed via dehydration (i.e., removal of H_2O).

Therefore, a basic anhydride is a base without water.

104. C is correct.

The problem is asking for the K_a of an acid, indicating that the acid in question is weak.

In aqueous solutions, a hypothetical weak acid HX partly dissociates and create this equilibrium:

$$HX \leftrightarrow H^+ + X^-$$

with an acid equilibrium constant, or

$$K_a = [H^+]\cdot[X^-] / [HX]$$

continued...

To calculate K_a, the concentration of all species is needed.

From the given pH, the concentration of H^+ ions can be calculated:

$[H^+] = 10^{-pH}$

$[H^+] = 10^{-7}$ M

The number of H^+ and X^- ions are equal; the concentration of X^- ions is also 10^{-7} M.

Those ions came from the dissociated acid molecules. According to the problem, only 24% of the acid is dissociated.

Therefore, the rest of acid molecules (100% – 24% = 76%) did not dissociate. Use the simple proportion of percentages to calculate the concentration of HX:

$[HX] = (76\% / 24\%) \times 1 \times 10^{-7}$ M

$[HX] = 3.17 \times 10^{-7}$ M

Use the concentration values to calculate K_a:

$K_a = [H^+]\cdot[X^-] / [HX]$

$K_a = [(1 \times 10^{-7}) \times (1 \times 10^{-7})] / (3.17 \times 10^{-7})$

$K_a = 3.16 \times 10^{-8}$

Calculate pK_a:

$pK_a = -\log K_a$

$pK_a = -\log (3.16 \times 10^{-8})$

$pK_a = 7.5$

105. E is correct.

The cations and anions on reactants switch pairs with a neutralization rxn, resulting in salt and water.

106. D is correct.

The strongest acid has the largest K_a value (or the lowest pK_a).

107. E is correct.

If the $[H_3O^+]$ is greater than 1×10^{-7}, it is an acidic solution.

If the $[H_3O^+]$ is less than 1×10^{-7}, it is a basic solution.

If the $[H_3O^+]$ equals 1×10^{-7}, it has a pH of 7, and the solution is neutral.

108. A is correct.

An electrolyte is a substance that dissociates into cations (i.e., positive ions) and anions (i.e., negative ions) when placed in solution. An electrolyte produces an electrically conducting solution when dissolved in a polar solvent (e.g., water). The dissolved ions disperse uniformly through the solvent.

If an electrical potential (i.e., the voltage generated by the battery) is applied to such a solution, the cations of the solution migrate towards the electrode (i.e., an abundance of electrons).

The anions migrate towards the electrode (i.e., a deficit of electrons).

109. D is correct.

$$H_3PO_4 \rightarrow H_2PO_4^- \rightarrow HPO_4^{2-}$$

In the Arrhenius theory, acids are substances that dissociate in an aqueous solution to produce H^+ (hydrogen ions).

In the Arrhenius theory, bases are substances that dissociate in an aqueous solution to produce ^-OH (hydroxide ions).

In the Brønsted–Lowry theory, acids and bases are defined by how they react.

The definition is expressed in terms of an equilibrium expression:

acid + base ↔ conjugate base + conjugate acid.

With an acid, HA, the equation can be written symbolically as:

$$HA + B \leftrightarrow A^- + HB^+$$

110. E is correct.

The Arrhenius acid–base theory states that acids produce H^+ ions in H_2O solution and bases produce OH^- ions in H_2O solution.

The Arrhenius acid–base theory states that neutralization happens when acid–base reactions produce water and salt and that these reactions must occur in an aqueous solution.

There are additional ways to classify acids and bases.

The Brønsted-Lowry acid–base theory focuses on the ability to accept and donate protons.

The Lewis acid–base theory focuses on accepting and donating electrons.

111. A is correct.

The Brønsted-Lowry acid–base theory focuses on the ability to accept and donate protons (H^+).

A Brønsted-Lowry acid is a term for a substance that donates a proton in an acid–base reaction, while a Brønsted-Lowry base is a substance that accepts a proton.

The Arrhenius acid–base theory focuses on producing H^+ and ^-OH ions.The Arrhenius acid–base theory states that neutralization happens when acid–base reactions produce water and salt and that these reactions must occur in an aqueous solution.

The Lewis acid–base theory focuses on accepting and donating electrons.

112. B is correct. By the Brønsted-Lowry acid–base theory:

An acid (reactant) dissociates a proton to become the conjugate base (product).

A base (reactant) gains a proton to become the conjugate acid (product).

The definition is expressed in terms of an equilibrium expression:

acid + base ↔ conjugate base + conjugate acid.

113. C is correct.

Water is neutral and has an equal concentration of hydroxide (^-OH) and hydronium (H_3O^+) ions.

114. D is correct.

The pI (isoelectric point) for an amino acid is defined as the pH for which an ionizable molecule has a net charge of zero.

In general, the net charge on the molecule is affected by pH, as it can become more positively or negatively charged due to the gain or loss of protons (H^+), respectively.

An amphoteric compound can react as an acid (i.e., donates protons) as well as a base (i.e., accepts protons).

A *zwitterion* is a molecule with a positive (cation) and negative (anion) region within the same molecule.

A zwitterion is *amphoteric*.

Amino acids (amino and carboxyl end) are an example of amphoteric molecules.

115. B is correct.

Spectator ions exist as a reactant and product in a chemical equation. A net ionic equation ignores the spectator ions that were part of the original equation.

A zwitterion is a molecule with a positive (cation) and negative (anion) region within the same molecule. A zwitterion is amphoteric. Amino acids (amino and carboxyl end) are an example of amphoteric molecules.

116. A is correct.

Salts that result from the reaction of strong acids and strong bases are neutral.

For example:

HCl	+	$NaOH$	$\leftrightarrow$	$NaCl + H_2O$
strong acid		strong base		neutral

117. D is correct.

Polyprotic means two H^+ that can dissociate.

118. E is correct.

An acid dissociates a proton to form the conjugate base, while a conjugate base accepts a proton to form the acid.

119. C is correct.

The products of a neutralization reaction (e.g., salt and water) are not corrosive.

NaOH and HCl are very corrosive.

However, after a neutralization reaction, they form NaCl, or common table salt, which is not corrosive.

120. C is correct.

Start by calculating the moles of $CaCO_3$:

Moles of $CaCO_3$ = mass of $CaCO_3$ / molecular mass of $CaCO_3$

Moles of $CaCO_3$ = 0.5 g / 100.09 g/mol

Moles of $CaCO_3$ = 0.005 mol

Use the coefficients in the reaction equation to find moles of HNO_3:

Moles of HNO_3 = (coefficient of HNO_3 / coefficient of $CaCO_3$) × moles of $CaCO_3$

Moles of HNO_3 = (2/1) × 0.005 mol

Moles of HNO_3 = 0.01 mol

Divide moles by volume to calculate molarity:

Molarity of HNO_3 = moles of HNO_3 / volume of HNO_3

Molarity of HNO_3 = 0.01 mol / (25 mL × 0.001 L/mL)

Molarity of HNO_3 = 0.4 M

Notes for active learning

Notes for active learning

Notes for active learning

Notes for active learning

APPENDIX

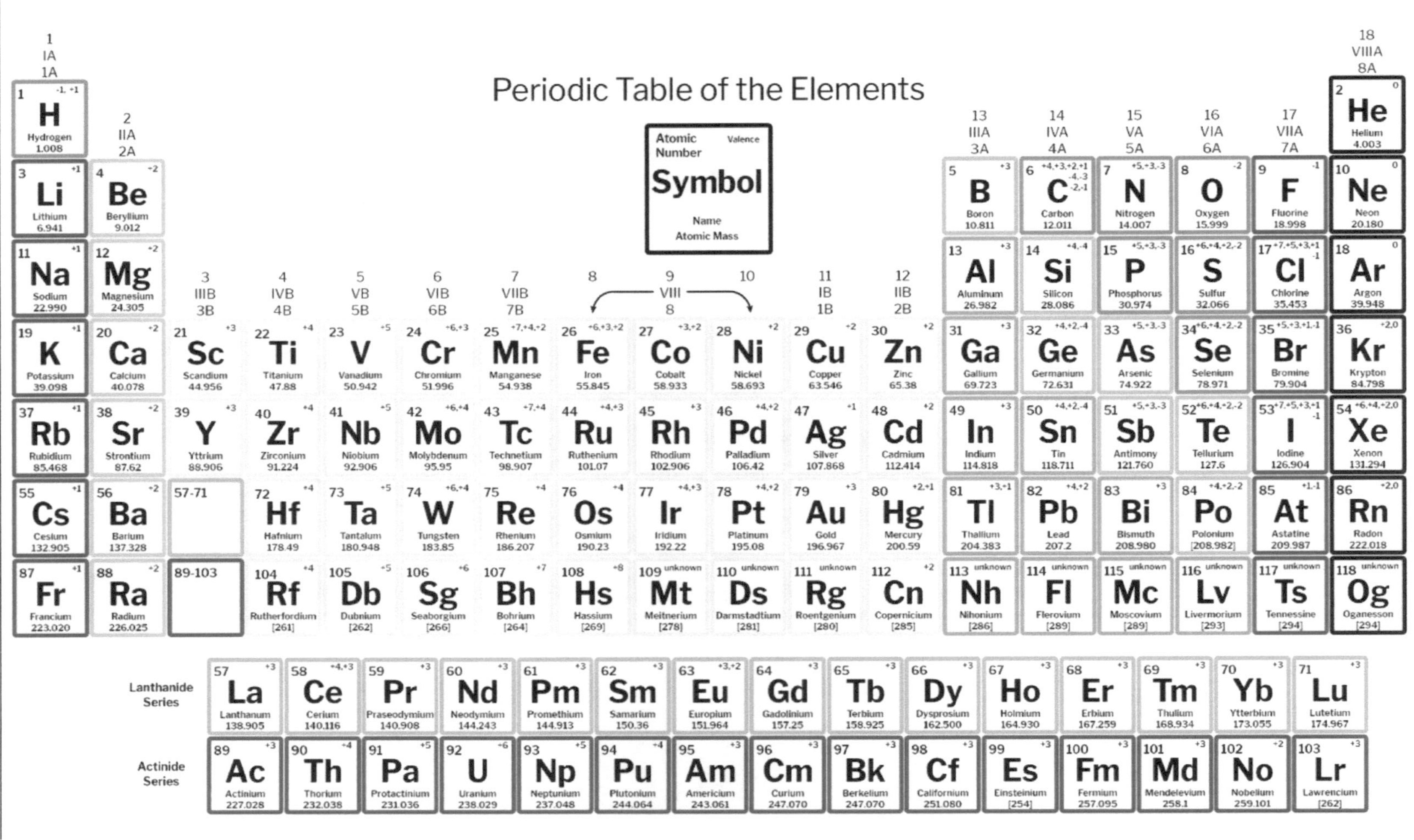
Periodic Table of the Elements
Atomic Number
Valence
Symbol
Name
Atomic Mass
1 IA 1A
2 IIA 2A
3 IIIB 3B
4 IVB 4B
5 VB 5B
6 VIB 6B
7 VIIB 7B
8
9 VIII 8
10
11 IB 1B
12 IIB 2B
13 IIIA 3A
14 IVA 4A
15 VA 5A
16 VIA 6A
17 VIIA 7A
18 VIIIA 8A
1 H Hydrogen 1.008
2 He Helium 4.003
3 Li Lithium 6.941
4 Be Beryllium 9.012
5 B Boron 10.811
6 C Carbon 12.011
7 N Nitrogen 14.007
8 O Oxygen 15.999
9 F Fluorine 18.998
10 Ne Neon 20.180
11 Na Sodium 22.990
12 Mg Magnesium 24.305
13 Al Aluminum 26.982
14 Si Silicon 28.086
15 P Phosphorus 30.974
16 S Sulfur 32.066
17 Cl Chlorine 35.453
18 Ar Argon 39.948
19 K Potassium 39.098
20 Ca Calcium 40.078
21 Sc Scandium 44.956
22 Ti Titanium 47.88
23 V Vanadium 50.942
24 Cr Chromium 51.996
25 Mn Manganese 54.938
26 Fe Iron 55.845
27 Co Cobalt 58.933
28 Ni Nickel 58.693
29 Cu Copper 63.546
30 Zn Zinc 65.38
31 Ga Gallium 69.723
32 Ge Germanium 72.631
33 As Arsenic 74.922
34 Se Selenium 78.971
35 Br Bromine 79.904
36 Kr Krypton 84.798
37 Rb Rubidium 85.468
38 Sr Strontium 87.62
39 Y Yttrium 88.906
40 Zr Zirconium 91.224
41 Nb Niobium 92.906
42 Mo Molybdenum 95.95
43 Tc Technetium 98.907
44 Ru Ruthenium 101.07
45 Rh Rhodium 102.906
46 Pd Palladium 106.42
47 Ag Silver 107.868
48 Cd Cadmium 112.414
49 In Indium 114.818
50 Sn Tin 118.711
51 Sb Antimony 121.760
52 Te Tellurium 127.6
53 I Iodine 126.904
54 Xe Xenon 131.294
55 Cs Cesium 132.905
56 Ba Barium 137.328
57-71
72 Hf Hafnium 178.49
73 Ta Tantalum 180.948
74 W Tungsten 183.85
75 Re Rhenium 186.207
76 Os Osmium 190.23
77 Ir Iridium 192.22
78 Pt Platinum 195.08
79 Au Gold 196.967
80 Hg Mercury 200.59
81 Tl Thallium 204.383
82 Pb Lead 207.2
83 Bi Bismuth 208.980
84 Po Polonium [208.982]
85 At Astatine 209.987
86 Rn Radon 222.018
87 Fr Francium 223.020
88 Ra Radium 226.025
89-103
104 Rf Rutherfordium [261]
105 Db Dubnium [262]
106 Sg Seaborgium [266]
107 Bh Bohrium [264]
108 Hs Hassium [269]
109 Mt Meitnerium [278]
110 Ds Darmstadtium [281]
111 Rg Roentgenium [280]
112 Cn Copernicium [285]
113 Nh Nihonium [286]
114 Fl Flerovium [289]
115 Mc Moscovium [289]
116 Lv Livermorium [293]
117 Ts Tennessine [294]
118 Og Oganesson [294]
Lanthanide Series
57 La Lanthanum 138.905
58 Ce Cerium 140.116
59 Pr Praseodymium 140.908
60 Nd Neodymium 144.243
61 Pm Promethium 144.913
62 Sm Samarium 150.36
63 Eu Europium 151.964
64 Gd Gadolinium 157.25
65 Tb Terbium 158.925
66 Dy Dysprosium 162.500
67 Ho Holmium 164.930
68 Er Erbium 167.259
69 Tm Thulium 168.934
70 Yb Ytterbium 173.055
71 Lu Lutetium 174.967
Actinide Series
89 Ac Actinium 227.028
90 Th Thorium 232.038
91 Pa Protactinium 231.036
92 U Uranium 238.029
93 Np Neptunium 237.048
94 Pu Plutonium 244.064
95 Am Americium 243.061
96 Cm Curium 247.070
97 Bk Berkelium 247.070
98 Cf Californium 251.080
99 Es Einsteinium [254]
100 Fm Fermium 257.095
101 Md Mendelevium 258.1
102 No Nobelium 259.101
103 Lr Lawrencium [262]

Notes for active learning

Common Chemistry Equations

Throughout the test the following symbols have the definitions specified unless otherwise noted.

L, mL	= liter(s), milliliter(s)	mm Hg	= millimeters of mercury
g	= gram(s)	J, kJ	= joule(s), kilojoule(s)
nm	= nanometer(s)	V	= volt(s)
atm	= atmosphere(s)	mol	= mole(s)

ATOMIC STRUCTURE

$E = h\nu$

$c = \lambda\nu$

E = energy
ν = frequency
λ = wavelength

Planck's constant, $h = 6.626 \times 10^{-34}$ J s

Speed of light, $c = 2.998 \times 10^{8}$ m s^{-1}

Avogadro's number $= 6.022 \times 10^{23}$ mol^{-1}

Electron charge, $e = -1.602 \times 10^{-19}$ coulomb

EQUILIBRIUM

$$K_c = \frac{[C]^c[D]^d}{[A]^a[B]^b}, \text{ where } a\,A + b\,B \rightleftarrows c\,C + d\,D$$

$$K_p = \frac{(P_C)^c(P_D)^d}{(P_A)^a(P_B)^b}$$

$$K_a = \frac{[H^+][A^-]}{[HA]}$$

$$K_b = \frac{[OH^-][HB^+]}{[B]}$$

$$K_w = [H^+][OH^-] = 1.0 \times 10^{-14} \text{ at } 25°C$$

$$= K_a \times K_b$$

$$pH = -\log[H^+],\ pOH = -\log[OH^-]$$

$$14 = pH + pOH$$

$$pH = pK_a + \log\frac{[A^-]}{[HA]}$$

$$pK_a = -\log K_a,\ pK_b = -\log K_b$$

Equilibrium Constants

K_c (molar concentrations)
K_p (gas pressures)
K_a (weak acid)
K_b (weak base)
K_w (water)

KINETICS

$$\ln[A]_t - \ln[A]_0 = -kt$$

$$\frac{1}{[A]_t} - \frac{1}{[A]_0} = kt$$

$$t_{1/2} = \frac{0.693}{k}$$

k = rate constant
t = time
$t_{1/2}$ = half-life

GASES, LIQUIDS, AND SOLUTIONS

$$PV = nRT$$

$$P_A = P_{total} \times X_A\text{, where } X_A = \frac{\text{moles A}}{\text{total moles}}$$

$$P_{total} = P_A + P_B + P_C + \ldots$$

$$n = \frac{m}{M}$$

$$K = {}^\circ C + 273$$

$$D = \frac{m}{V}$$

$$KE \text{ per molecule} = \frac{1}{2}mv^2$$

Molarity, M = moles of solute per liter of solution

$$A = abc$$

P = pressure
V = volume
T = temperature
n = number of moles
m = mass
M = molar mass
D = density
KE = kinetic energy
v = velocity
A = absorbance
a = molar absorptivity
b = path length
c = concentration

Gas constant, $R = 8.314 \text{ J mol}^{-1}\text{ K}^{-1}$
$= 0.08206 \text{ L atm mol}^{-1}\text{ K}^{-1}$
$= 62.36 \text{ L torr mol}^{-1}\text{ K}^{-1}$
1 atm = 760 mm Hg
= 760 torr
STP = 0.00°C and 10^5 Pa

THERMOCHEMISTRY/ ELECTROCHEMISTRY

$$q = mc\Delta T$$

$$\Delta S^\circ = \sum S^\circ \text{ products} - \sum S^\circ \text{ reactants}$$

$$\Delta H^\circ = \sum \Delta H_f^\circ \text{ products} - \sum \Delta H_f^\circ \text{ reactants}$$

$$\Delta G^\circ = \sum \Delta G_f^\circ \text{ products} - \sum \Delta G_f^\circ \text{ reactants}$$

$$\Delta G^\circ = \Delta H^\circ - T\Delta S^\circ$$
$$= -RT \ln K$$
$$= -nFE^\circ$$

$$I = \frac{q}{t}$$

q = heat
m = mass
c = specific heat capacity
T = temperature
S° = standard entropy
H° = standard enthalpy
G° = standard free energy
n = number of moles
E° = standard reduction potential
I = current (amperes)
q = charge (coulombs)
t = time (seconds)

Faraday's constant, F = 96,485 coulombs per mole of electrons

$$1 \text{ volt} = \frac{1 \text{ joule}}{1 \text{ coulomb}}$$

Glossary of Chemistry Terms

A

Absolute entropy (of a substance) – the increase in the entropy of a substance as it goes from a perfectly ordered crystalline form at 0 K (where its entropy is zero) to the temperature in question.

Absolute zero – the zero point on the absolute temperature scale; –273.15 °C or 0 K; theoretically, the temperature at which molecular motion ceases (i.e., the system does not emit or absorb energy, atoms at rest).

Absorption spectrum – spectrum associated with absorption of electromagnetic radiation by atoms (or other species), resulting from transitions from lower to higher energy states.

Accuracy – how close a value is to the actual value; see *precision.*

Acid – a substance that produces H^+ *(aq)* ions in an aqueous solution and gives a pH of less than 7.0; strong acids ionize entirely or almost entirely in dilute aqueous solution; weak acids ionize only slightly. It turns litmus red.

Acid dissociation constant – an equilibrium constant for dissociating a weak acid.

Acid rain – rainwater with a pH of less than 5.7; caused by the gases NO_2 (vehicle exhaust fumes) and SO_2 (from burning fossil fuels) dissolving in the rain. It kills fish, wildlife, and trees and destroys buildings and lakes.

Acidic salt – contains an ionizable hydrogen atom; does not necessarily produce acidic solutions.

Actinides – the fifteen chemical elements that are between actinium (89) and lawrencium (103).

Activated complex – a structure forming because of a collision between molecules while new bonds form.

Activation energy – the amount of energy that reactants must absorb in their ground states to reach the transition state needed for a reaction can occur.

Active metal – a metal with low ionization energy that loses electrons readily to form cations.

Activity (of a component of ideal mixture) – a dimensionless quantity whose magnitude is equal to the molar concentration in an ideal solution; equal to partial pressure in an ideal gas mixture; 1 for pure solids or liquids.

Activity series – a listing of metals (and hydrogen) in order of decreasing activity.

Actual yield – the amount of a specified pure product obtained from a given reaction; see *theoretical yield.*

Addition reaction – a reaction in which two atoms or groups of atoms are added to a molecule, one on each side of a double or triple bond.

Adhesive forces – forces of attraction between a liquid and another surface.

Adsorption – the adhesion of a species onto the surfaces of particles.

Aeration – the mixing of air into a liquid or a solid.

Alcohol – hydrocarbon derivative containing a ~OH group attached to a carbon atom, not in an aromatic ring.

Alkali metals – the elements of Group IA on the periodic table (e.g., Na, K, Rb).

Alkaline battery – a dry cell in which the electrolyte contains KOH.

Alkaline earth metals – group IIA metals on the periodic table; see *earth metals*.

Allomer – a substance that has a different composition than another but the same crystalline structure.

Allotropes – elements with different structures (therefore different forms), such as carbon (e.g., diamonds, graphite, and fullerene).

Allotropic modifications (allotropes) – different forms of the same element in the same physical state.

Alloy – a mixture of metals. For example, bronze is an alloy formed from copper and tin.

Alloying – mixing metal with other substances (usually other metals) to modify its properties.

Alpha (α) particle – a helium nucleus; helium ion with 2+ charge; an assembly of two protons and two neutrons.

Amorphous solid – a non-crystalline solid with no well-defined ordered structure.

Ampere – unit of electrical current; one ampere equals one coulomb per second.

Amphiprotism – the ability of a substance to exhibit amphiprotic by accepting donated protons.

Amphoterism – the ability to react with both acids and bases; to act as either an acid or a base.

Amplitude – the maximum distance that medium particles carrying the wave move from their rest position.

Anion – a negative ion; an atom or group of atoms that has gained one or more electrons.

Anode – in a cathode ray tube, the positive electrode (electrode at which oxidation occurs); the positive side of a dry cell battery or a cell.

Antibonding orbital – a molecular orbital higher in energy than any of the atomic orbitals from which it is derived; lends instability to a molecule or ion when populated with electrons; denoted with star (*) superscript.

Artificial transmutation – an artificially induced nuclear reaction caused by the bombardment of a nucleus with subatomic particles or small nuclei.

Associated ions – short-lived species formed by the collision of dissolved ions of opposite charges.

Atmosphere – a unit of pressure; the pressure supports a column of mercury 760 mm high at 0 °C.

Atom – a chemical element in its smallest form; it comprises neutrons and protons within the nucleus and electrons circling the nucleus; it is the smallest part of an element that can exist.

Atomic mass unit (amu) – one-twelfth of the mass of an atom of the carbon-12 isotope; used for stating atomic and formula weights; known as a dalton.

Atomic number – represents an element corresponding with the number of protons within the nucleus; the number of protons in the nucleus of the atom.

Atomic orbital (*AO*) – a region or volume in space where the probability of finding electrons is highest.

Atomic radius – radius of an atom.

Atomic weight – weighted average of the masses of the constituent isotopes of an element; the relative masses of atoms of different elements.

Aufbau (or *building up*) principle – describes the order in which electrons fill orbitals in atoms.

Autoionization – an ionization reaction between identical molecules.

Avogadro's Law – equal volumes of gases contain the same number of molecules at the same temperature and pressure.

Avogadro's number (N_A) – the number (6.022×10^{23}) of atoms, molecules, or particles found in precisely 1 mole of a substance.

B

Background radiation – extraneous to an experiment; usually the low-level natural radiation from cosmic rays and trace radioactive substances present in the environment.

Band – a series of very closely spaced nearly continuous molecular orbitals that belong to the crystal as a whole.

Band of stability – band containing nonradioactive nuclides in a plot of neutrons *vs.* their atomic number.

Band theory of metals – the theory that accounts for the bonding and properties of metallic solids.

Barometer – a device used to measure the pressure in the atmosphere.

Base – a substance that produces ^{-}OH (*aq*) ions in an aqueous solution; accepts a proton and has a high pH; strongly soluble bases are soluble in water and are entirely dissociated; weak bases ionize only slightly; a typical example of a base is sodium hydroxide (NaOH). It turns litmus blue.

Basic anhydride – the oxide of a metal that reacts with water to form a base.

Basic salt – a salt containing an ionizable OH group.

Beta (β) particle – an electron emitted from the nucleus when a neutron decays to a proton and an electron.

Binary acid – a binary compound in which H is bonded to one or more electronegative nonmetals.

Binary compound – consists of two elements; it may be ionic or covalent.

Binding energy (nuclear binding energy) – the energy equivalent ($E = mc^2$) of the mass deficiency of an atom (where E is the energy in joules, m is the mass in kilograms, and c is the speed of light in m/s^2).

Boiling – the phase transition of liquid vaporizing.

Boiling point – the temperature at which the vapor pressure of a liquid is equal to the applied pressure; the *condensation point*.

Boiling point elevation – the increase in the boiling point of a solvent caused by the dissolution of a nonvolatile solute.

Bomb calorimeter – a device used to measure the heat transfer between a system and its surroundings at constant volume.

Bond – the attraction and repulsion between atoms and molecules is a cornerstone of chemistry.

Bond energy – the amount of energy necessary to break one mole of bonds in a substance, dissociating the substance in its gaseous state into atoms of its elements in the gaseous state.

Bond order – half the number of electrons in bonding orbitals minus half the electrons in antibonding orbitals.

Bonding orbital – a molecular orbit lower in energy than any of the atomic orbitals from which it is derived; lends stability to a molecule or ion when populated with electrons.

Bonding pair – pair of electrons involved in a covalent bond.

Boron hydrides – binary compounds of boron and hydrogen.

Born-Haber cycle – a series of reactions (and the accompanying enthalpy changes) which, when summed, represents the hypothetical one-step reaction by which elements in their standard states are converted into crystals of ionic compounds (and the accompanying enthalpy changes).

Boyle's Law – at a constant temperature, the volume occupied by a definite mass of a gas is inversely proportional to the applied pressure.

Breeder reactor – a nuclear reactor that produces more fissionable nuclear fuel than it consumes.

Brønsted-Lowrey acid – a chemical species that donates a proton.

Brønsted-Lowrey base – a chemical species that accepts a proton.

Buffer solution – resists change in pH; contains either a weak acid and a soluble ionic salt of the acid or a weak base and a soluble ionic salt of the base.

Buret – a piece of volumetric glassware, usually graduated in 0.1 mL intervals, used to deliver solutions for titrations in a quantitative (drop-like) manner; also spelled *burette.*

C

Calorie – the amount of heat required to raise the temperature of one gram of water from 14.5 °C to 15.5 °C; 1 calorie = 4.184 joules.

Calorimeter – a device used to measure the heat transfer between a system and its surroundings.

Canal ray – a stream of positively charged particles (cations) that moves toward the negative electrode in cathode ray tubes; observed to pass through canals in the negative electrode.

Capillary – a tube having a very small inside diameter.

Capillary action – the drawing of a liquid up the inside of a small-bore tube when adhesive forces exceed cohesive forces; the depression of the surface of the liquid when cohesive forces exceed the adhesive forces.

Catalyst – a chemical compound used to change the rate (to speed or slow it) of a regenerated reaction (i.e., not consumed) at the end of the reaction.

Catenation – the bonding of atoms of the same element into chains or rings (i.e., the ability of an element to bond with itself).

Cathode – the electrode at which reduction occurs; in a cathode ray tube, the negative electrode.

Cathodic protection – protection of a metal (making a cathode) against corrosion by attaching it to a sacrificial anode of more easily oxidized metal.

Cathode ray tube – a closed glass tube containing gas under low pressure, with electrodes near the ends and a luminescent screen near the positive electrode; produces cathode rays when a high voltage is applied.

Cation – a positive ion; an atom or group of atoms that lost one or more electrons.

Cell potential – the potential difference, E_{cell}, between oxidation and reduction half-cells under nonstandard conditions; the force in a galvanic cell pulls electrons through a reducing agent to an oxidizing agent.

Central atom – an atom in a molecule or polyatomic ion bonded to more than one other atom.

Chain reaction – a reaction that, once initiated, sustains itself and expands; a reaction in which reactive species, such as radicals, are produced in more than one step; these reactive species propagate the chain reaction.

Charles' Law – at constant pressure, the volume occupied by a definite mass of gas is directly proportional to its absolute temperature.

Chemical bonds – the attractive forces holding atoms together in elements or compounds.

Chemical change – when one or more new substances are formed.

Chemical equation – description of a chemical reaction by placing the formulas of the reactants on the left of an arrow and the formulas of the products on the right.

Chemical equilibrium – a state of dynamic balance in which the rates of forward and reverse reactions are equal; there is no net change in concentrations of reactants or products while a system is at equilibrium.

Chemical kinetics – the study of rates and mechanisms of chemical reactions and factors they depend on.

Chemical periodicity – the variations in properties of elements with their position in the periodic table.

Chemical reaction – the change of one or more substances into another or multiple substances.

Cloud chamber – a device for observing the paths of speeding particles as vapor molecules condense on them to form fog-like tracks.

Cobalt chloride paper – water test; water changes the color from blue to pink.

Coefficient of expansion – the ratio of the change in the length or the volume of a body to the original length or volume for a unit change in temperature.

Cohesive forces – the forces of attraction among particles of a liquid.

Colligative properties – physical properties of solutions that depend upon the number but not the kind of solute particles present.

Collision theory – theory of reaction rates that states that effective collisions between reactant molecules must occur for the reaction to occur.

Colloid – a heterogeneous mixture in which solute-like particles do not settle out (e.g., many kinds of milk).

Combination reaction – two substances (elements or compounds) combine to form one compound.

Combustible – classification of liquid substances that burn based on flashpoints; any liquid having a flashpoint at or above 37.8 °C (100 °F) but below 93.3 °C (200 °F), except any mixture having components with flashpoints of 93.3 °C (200 °F) or higher, the total of which makes up 99% or more of the volume of the mixture.

Combustion (or *burning*) – an exothermic reaction between an oxidant and fuel with heat and often light.

Common ion effect – suppression of ionization of a weak electrolyte by the presence in the same solution of a strong electrolyte containing one of the same ions as the weak electrolyte.

Complex ions – ions resulting from coordinating covalent bonds between simple ions and other ions or molecules.

Composition stoichiometry – describes the quantitative (mass) relationships among elements in compounds.

Compound – a substance of two or more chemically bonded elements in fixed proportions; can be decomposed into constituent elements.

Compressed gas – a single or mixture of gases having (in a container) an absolute pressure exceeding 40 psi at 21.1 °C (70 °F).

Compression – an area in a longitudinal wave where the particles are closer and pushed in.

Concentration – the amount of solute per unit volume, the mass of solvent or solution.

Condensation – the phase change from gas to liquid.

Condensed phases – the liquid and solid phases; phases in which particles interact strongly.

Condensed states – the solid and liquid states.

Conduction – heat transfer between substances in direct contact with each other (i.e., must be touching); when particles of a hotter substance vibrate, these molecules bump into nearby particles and transfer some energy.

Conduction band – a vacant or partially filled band of energy levels just higher in energy than a filled band; a band within which, or into which, electrons must be promoted to allow electrical conduction to occur in a solid.

Conductor – material that allows electric flow more freely.

Conjugate acid-base pair – in Brønsted-Lowry terms, a reactant and a product that differ by a proton, H^+.

Conformations – structures of a compound that differ by the extent of their rotation about a single bond.

Continuous spectrum – contains wavelengths in a specified region of the electromagnetic spectrum.

Control rods – rods of materials such as cadmium or boron steel that act as neutron absorbers (not merely moderators), used in nuclear reactors to control neutron fluxes and therefore fission rates.

Conjugated double bonds – double bonds separated from each other by one single bond –C=C–C=C–

Contact process – the industrial process for sulfur trioxide and sulfuric acid production from sulfur dioxide.

Convection – the physical flow of matter when heat flows by energized molecules from one place to another through the movement of fluids. The transfer of heat through a liquid or a gas when molecules of the liquid or gas move and carry the heat.

Coordinate covalent bond – a covalent bond with shared electrons furnished by the same species; a bond between a Lewis acid and a Lewis base.

Coordination compound or complex – a compound containing coordinate covalent bonds.

Coordination number – the number of donor atoms coordinated to metal; the number of nearest neighbors of an atom or ion in describing crystals.

Coordination sphere – the metal ion and its coordinating ligands but no uncoordinated counter-ions.

Corrosion – oxidation of metals (e.g., rusting) in the presence of air and moisture.

Coulomb – the SI unit of electrical charge; unit symbol – C.

Covalent bond – a force of attraction (chemical bond) formed by the sharing of electron pairs between two atoms.

Covalent compounds – compounds made of two or more nonmetal atoms bonded by sharing valence electrons.

Critical mass – the minimum mass of a particular fissionable nuclide in a given volume required to sustain a nuclear chain reaction.

Critical point – the combination of critical temperature and critical pressure of a substance.

Critical pressure – the pressure required to liquefy a gas (vapor) at its *Critical temperature*.

Critical temperature – the temperature above which a gas cannot be liquefied; the temperature above which a substance cannot exhibit distinct gas and liquid phases.

Crystal – a solid packed with ions, molecules, or atoms in an orderly fashion.

Crystal field stabilization energy – a measure of the net energy of stabilization gained by a metal ion's nonbonding d electrons due to complex formation.

Crystal field theory – bonding in transition metal complexes in which ligands and metal ions are treated as point charges; a purely ionic model; ligand point charges represent the crystal (electrical) field perturbing the metal's d orbitals containing nonbonding electrons.

Crystal lattice – a pattern of arrangement of particles in a crystal.

Crystal lattice energy – the amount of energy that holds a crystal together; the energy change when a mole of solid forms fom its constituent molecules or ions (for ionic compounds) in their gaseous state (always negative).

Crystalline solid – a solid characterized by a regular, ordered arrangement of particles.

Curie (Ci) – the basic unit to describe the intensity of radioactivity in a sample of material; one curie equals 37 billion disintegrations per second or approximately the amount of radioactivity given off by 1 gram of radium.

Current – a flow of charged particles, such as electrons or ions, moving through an electrical space or conductor. It is measured as the net rate of flow of electric charge; the unit is Ampere (A).

Cuvette – glassware used in spectroscopic experiments; usually made of plastic, glass, or quartz and should be as clean and transparent as possible.

Cyclotron – a device for accelerating charged particles along a spiral path.

D

Dalton's Law (or the *law of partial pressures*) – the pressure exerted by a mixture of gases is the sum of the partial pressures of the individual gases.

Daughter nuclide – nuclide produced in nuclear decay.

Debye – the unit used to express dipole moments.

Degenerate – in orbitals, describes orbitals of the same energy.

Deionization – the removal of ions; in the case of water, mineral ions such as sodium, iron and calcium.

Deliquescence – substances that absorb water from the atmosphere to form liquid solutions.

Delocalization – in reference to electrons, bonding electrons distributed among more than two atoms bonded; occurs in species that exhibit resonance.

Density – mass per unit volume; D = M × V.

Deposition – settling particles within a solution; the direct solidification of vapor by cooling; see *sublimation*.

Derivative – a compound that can be imagined arising from a parent compound by replacing one atom with another atom or group of atoms; used extensively in organic chemistry to identify compounds.

Detergent – a soap-like emulsifier with a sulfate, SO_3, or a phosphate group instead of a carboxylate group.

Deuterium – an isotope of hydrogen whose atoms are twice as massive as ordinary hydrogen; deuterium atoms contain a proton and a neutron in the nucleus.

Dextrorotatory – refers to an optically active substance that rotates plane-polarized light clockwise, also known as *dextro*.

Diagonal similarities – chemical similarities in the Periodic Table of Elements of Period 2 to elements of Period 3 one group to the right, especially evident toward the left of the periodic table.

Diamagnetism – weak repulsion by a magnetic field.

Differential Scanning Calorimetry (DSC) – a technique for measuring temperature, direction, and magnitude of thermal transitions in a sample material by heating/cooling and comparing the amount of energy required to maintain its rate of temperature increase or decrease with an inert reference material under similar conditions.

Differential Thermal Analysis (DTA) – a technique for observing the temperature, direction and magnitude of thermally induced transitions in a material by heating/cooling a sample and comparing its temperature with an inert reference material under similar conditions.

Differential thermometer – a thermometer used to measure very small temperature changes accurately.

Dilution – the process of reducing the concentration of a solute in a solution, usually by mixing with more solvent.

Dimer – molecule formed by combining two smaller (identical) molecules.

Dipole – electric or magnetic separation of charge; charge separation between two covalently bonded atoms.

Dipole-dipole interactions – attractive electrostatic forces between polar molecules (i.e., between molecules with permanent dipoles).

Dipole moment – the product of the distance separating opposite charges of an equal magnitude of charge; a measure of the polarity of a bond or molecule; a measured dipole refers to the dipole moment of an entire molecule.

Dispersing medium – the solvent-like phase in a colloid.

Dispersed phase – the solute-like species in a colloid.

Displacement reactions – reactions in which one element displaces another from a compound.

Disproportionation reactions – redox reactions in which the oxidizing agent and the reducing agent are the same species.

Dissociation – in an aqueous solution, the process by which a solid ionic compound separates into its ions.

Dissociation constant – equilibrium constant for dissociating a complex ion into a simple ion and coordinating species (ligands).

Dissolution or solvation – the spread of ions in a monosaccharide.

Distilland – the material in a distillation apparatus that is to be distilled.

Distillate – the material in a distillation apparatus collected in the receiver.

Distillation – separating a liquid mixture into its components based on differences in boiling points; the process in which components of a mixture are separated by boiling away the more volatile liquid; the vaporization of a liquid by heating and then the condensation of the vapor by cooling.

Domain – a cluster of atoms in a ferromagnetic substance, which align in the same direction in the presence of an external magnetic field.

Donor atom – a ligand atom whose electrons are shared with a Lewis acid.

***d*-orbitals –** beginning in the third energy level, a set of five degenerate orbitals per energy level, higher in energy than s and p orbitals of the same energy level.

Dosimeter – a small, calibrated electroscope worn by laboratory personnel to measure incident ionizing radiation or chemical exposure.

Double bond – covalent bond resulting from the sharing of four electrons (two pairs) between two atoms.

Double salt – solid consisting of two co-crystallized salts.

Doublet – two peaks or bands of about equal intensity appearing close on a spectrogram.

Downs cell – electrolytic cell for the commercial electrolysis of molten sodium chloride.

DP number – the degree of polymerization; the average number of monomer units per polymer unit.

Dry cells (voltaic cells) – ordinary batteries for appliances (e.g., flashlights, radios).

Dumas method – a method used to determine the molecular weights of volatile liquids.

Dynamic equilibrium – an equilibrium in which the processes occur continuously with no net change.

E

Earth metal – highly reactive elements in group IIA of the periodic table (includes beryllium, magnesium, calcium, strontium, barium, and radium); see *alkaline earth metal.*

Effective collisions – a collision between molecules resulting in a reaction; one in which the molecules collide with proper relative orientations and sufficient energy to react.

Effective molality – the sum of the molalities of solute particles in a solution.

Effective nuclear charge – the nuclear charge experienced by the outermost electrons of an atom; the actual nuclear charge minus the effects of shielding due to inner-shell electrons (e.g., a set of dx_2-y_2 and dz_2 orbitals); those d orbitals within a set with lobes directed along the *x*, *y* and *z*-axes.

Electrical conductivity – the measure of how easily an electric current can flow through a substance.

Electric charge – a measured property (coulombs) that determines electromagnetic interaction.

Electrochemical cell – using a chemical reaction's current; electromotive force is made.

Electrochemistry – the study of chemical changes produced by electrical current and electricity production by chemical reactions.

Electrodes – surfaces upon which oxidation and reduction half-reactions occur in electrochemical cells; a conductor dips into an electrolyte and allows the electrons to flow to and from the electrolyte.

Electrode potentials – potentials, E, of half-reactions as reductions versus the standard hydrogen electrode.

Electrolysis – 1) occurs in electrolytic cells; chemical decomposition occurs by passing an electric current through a solution containing ions. 2) producing a chemical change using electricity; used to split up water into H and O_2.

Electrolyte – a substance (i.e., anions, cations) which, when dissolved in water, conducts electricity. An ionic solution that conducts a certain amount of current and split categorically as weak and strong.

Electrolytic cells – electrochemical cells in which electrical energy causes nonspontaneous redox reactions to occur (i.e., forced to occur by applying an outside source of electrical energy).

Electrolytic conduction – electrical current passes by ions through a solution or pure liquid.

Electromagnetic radiation – energy propagated using electric and magnetic fields that oscillate in directions perpendicular to the direction of travel of the energy; a type of wave that can go through vacuums as well as material; classified as a "self-propagating wave."

Electromagnetism – fields of an electric charge and electric properties that change how particles move and interact.

Electromotive force – a device that gains energy as electric charges pass through it.

Electromotive series – the relative order of tendencies for elements and their simple ions to act as oxidizing or reducing agents; also known as the "activity series."

Electron – a subatomic particle having a mass of 0.00054858 amu and a charge of 1−.

Electron affinity – the amount of energy absorbed in the process in which an electron is added to a neutral isolated gaseous atom to form a gaseous ion with a 1− charge; it has a negative value if energy is released.

Electron configuration – the specific distribution of electrons in atomic orbitals of atoms or ions.

Electron-deficient compounds –contain at least one atom (other than H) that shares fewer than eight electrons.

Electron shells – an orbital around the atom's nucleus with a fixed number of electrons (usually two or eight).

Electronic transition – the transfer of an electron from one energy level to another.

Electronegativity – a measure of the relative tendency of an atom to attract electrons to itself when chemically combined with another atom.

Electronic geometry – the geometric arrangement of orbitals containing the shared and unshared electron pairs surrounding the central atom or polyatomic ion.

Electrophile – positively charged or electron-deficient.

Electrophoresis – a technique for separating ions by migration rate and direction of migration in an electric field.

Electroplating – a metal is covered with another metal layer using electricity; plating a metal onto a (cathodic) surface by electrolysis.

Element – a substance that cannot be decomposed into simpler substances by chemical means; defined by its *atomic number*. A substance that cannot be split into simpler substances by chemical means.

Eluant (or eluent) – the solvent used in the process of elution, as in liquid chromatography.

Eluate – a solvent (or mobile phase) which passes through a chromatographic column and removes the sample components from the stationary phase.

Emission spectrum – the emission of electromagnetic radiation by atoms (or other species) resulting from electronic transitions from higher to lower energy states.

Empirical formula – gives the simplest whole-number ratio of atoms of each element present in a compound; also known as the simplest formula.

Emulsifying agent – a substance that coats the particles of the dispersed phase and prevents coagulation of colloidal particles; an emulsifier.

Emulsion – colloidal suspension of a liquid in a liquid.

Endothermic – describes processes that absorb heat energy (H).

Endothermicity – the absorption of heat by a system as the process occurs.

Endpoint – the point at which an indicator changes color and a titration is stopped.

Energy – a system's ability to do work.

Enthalpy (*H*) – the heat content of a specific amount of substance; E= PV.

Entropy (*S*) – a thermodynamic state or property that measures the degree of disorder (i.e., randomness) of a system; the amount of energy not available for work in a closed thermodynamic system (usually denoted by S).

Enzyme – a protein that acts as a catalyst in biological systems.

Equation of state – an expression describes the behavior of matter in a given state; the van der Waals equation describes the behavior of the gaseous state.

Equilibrium or chemical equilibrium – a state of dynamic balance with the rates of forward and reverse reactions equal; the state of a system when neither forward nor reverse reaction is thermodynamically favored.

Equilibrium constant – a quantity that characterizes the equilibrium position for a reversible reaction; its magnitude is equal to the mass action expression at equilibrium; equilibrium, "K," varies with temperature.

Equivalence point – the point when chemically equivalent amounts of reactants have reacted.

Equivalent weight – an oxidizing or reducing agent whose mass gains (oxidizing agents) or loses (reducing agents) 6.022×10^{23} electrons in a redox reaction.

Evaporation – vaporization of a liquid below its boiling point.

Evaporation rate – the rate at which a particular substance will vaporize (evaporate) compared to the rate of a known substance such as ethyl ether, especially useful for health and fire-hazard considerations.

Excited state – any state other than the ground state of an atom or molecule; see *ground state.*

Exothermic – describes processes that release heat energy (H).

Exothermicity – the release of heat by a system as a process occurs.

Explosive – a chemical or compound that causes a sudden, almost instantaneous release of pressure, gas, heat, and light when subjected to sudden shock, pressure, high temperature, or applied potential.

Explosive limits – the range of concentrations over which a flammable vapor mixed with the proper ratios of air will ignite or explode if a source of ignition is provided.

Extensive property – a property that depends upon the amount of material in a sample.

Extrapolate – to estimate the value of a result outside the range of a series of known values; a technique used in standard additions calibration procedure.

F

Faraday constant – a unit of electrical charge widely used in electrochemistry and equal to ~ 96,500 coulombs; represents 1 mole of electrons, or the Avogadro number of electrons: 6.022×10^{23} electrons.

Faraday's law of electrolysis – a two-part law that Michael Faraday published about electrolysis. 1. the mass of a substance altered at an electrode during electrolysis is directly proportional to the quantity of electricity transferred at that electrode. 2. the mass of an elemental material altered at an electrode is directly proportional to the element's equivalent weight; one equivalent weight of a substance is produced at each electrode during the passage of 96,487 coulombs of charge through an electrolytic cell.

Fast neutron – a neutron ejected at high kinetic energy in a nuclear reaction.

Ferromagnetism – the ability of a substance to become permanently magnetized by exposure to an external magnetic field.

Flashpoint – the temperature at which a liquid will yield enough flammable vapor to ignite; there are various recognized industrial testing methods; therefore, the method used must be stated.

Fluorescence – absorption of high energy radiation by a substance and subsequent emission of visible light.

First Law of Thermodynamics – the amount of energy in the universe is constant (i.e., energy is neither created nor destroyed in ordinary chemical reactions and physical changes); known as the Law of Conservation of Energy.

Fluids – substances that flow freely; gases and liquids.

Flux – a substance added to react with the charge or a product of its reduction; in metallurgy, it is usually added to lower a melting point.

Foam – colloidal suspension of a gas in a liquid.

Formal charge – a method of counting electrons in a covalently bonded molecule or ion; it counts bonding electrons as though they were equally shared between the two atoms.

Formula – a combination of symbols that indicates the chemical composition of a substance.

Formula unit – the smallest repeating unit of a substance; the molecule for nonionic substances.

Formula weight – the mass of one formula unit of a substance in atomic mass units.

Fossil fuels – formed from the remains of plants and animals that lived millions of years ago.

Fractional distillation – when a fractioning column is used in a distillation apparatus to separate the components of a liquid mixture with different boiling points.

Fractional precipitation – removal of some ions from a solution by precipitation while leaving other ions with similar properties in the solution.

Free energy change – the indicator of the spontaneity of a process at constant T and P (e.g., if ΔG is negative, the process is spontaneous).

Free radical – a highly reactive chemical species carrying no charge and having a single unpaired electron in an orbital.

Freezing – phase transition from liquid to solid.

Freezing point depression – the decrease in the freezing point of a solvent caused by the presence of a solute.

Frequency – the number of repeating points on a wave that passes a given observation point per unit time; the unit is 1 hertz = 1 cycle per 1 second.

Fuel – any substance that burns in oxygen to produce heat.

Fuel cells – a voltaic cell that converts the chemical energy of a fuel and an oxidizing agent directly into electrical energy continuously.

G

Gamma (γ) ray – a highly penetrating type of nuclear radiation similar to x-ray radiation, except that it comes from within the nucleus of an atom and has higher energy; energy-wise, very similar to cosmic rays except that cosmic rays originate from outer space.

Galvanic cell – battery made up of electrochemical with two different metals connected by a salt bridge.

Galvanizing – placing a thin layer of zinc on a ferrous material to protect the underlying surface from corrosion.

Gangue – sand, rock and other impurities surrounding the mineral of interest in an ore.

Gas – a state of matter in which the particles have no definite shape or volume, though they fill their container.

Gay-Lussac's Law – the expression used for each of the two relationships named after the French chemist Joseph Louis Gay-Lussac concerning the properties of gases; more usually applied to his law of combining volumes.

Geiger counter – a gas-filled tube that discharges electrically when ionizing radiation passes through it.

Gel – colloidal suspension of a solid dispersed in a liquid; a semi-rigid solid.

Gibbs (free) energy – the thermodynamic state function of a system that indicates the amount of energy available for the system to do useful work at constant T and P; a value that indicates the spontaneity of a reaction (usually denoted by G).

Graham's Law – the rates of effusion of gases are inversely proportional to the square roots of their molecular weights or densities.

Ground state – the lowest energy state or most stable state of an atom, molecule, or ion; see *excited state.*

Group – a vertical column in the periodic table; known as a family.

H

Haber process – a process for the catalyzed industrial production of ammonia from N_2 and H_2 at high temperature and pressure.

Half-cell – the compartment in which the oxidation or reduction half-reaction occurs in a voltaic cell.

Half-life – the time required for half of a reactant to be converted into product(s); the time required for half of a given sample to undergo radioactive decay.

Half-reaction – the oxidation or the reduction part of a redox reaction.

Halogens – group VIIA elements: F, Cl, Br, I; halogens are nonmetals.

Hard water – water high in dissolved minerals that is it difficult to form lather with soap.

Heat – a form of energy that flows between two samples of matter because of their temperature differences.

Heat capacity – the amount of heat required to raise the temperature of a mass one degree Celsius.

Heat of condensation – the amount of heat that must be removed from one gram of vapor at its condensation point to condense the vapor with no change in temperature.

Heat of crystallization – the amount of heat that must be removed from one gram of a liquid at its freezing point to freeze it with no change in temperature.

Heat of fusion – the amount of heat required to melt one gram of a solid at its melting point with no change in temperature; usually expressed in J/g; the molar heat of fusion is the amount of heat required to melt one mole of a solid at its melting point with no change in temperature and is usually expressed in kJ/mol.

Heat of solution – the amount of heat absorbed in forming a solution that contains one mole of the solute; the value is positive if heat is absorbed (endothermic) and negative if heat is released (exothermic).

Heat of vaporization – the amount of heat required to vaporize one gram of a liquid at its boiling point with no change in temperature; usually expressed in J/g; the molar heat of vaporization is the amount of heat required to vaporize one mole of liquid at its boiling point with no change in temperature and is usually expressed as ion kJ/mol.

Heisenberg uncertainty principle – states that it is impossible to accurately determine the *momentum* and *position* of an electron simultaneously.

Henry's Law – the gas pressure above a solution is proportional to the concentration of the gas in the solution.

Hess' Law of heat summation – the enthalpy change for a reaction is the same whether it occurs in one step or a series of steps.

Heterogeneous catalyst – exist in a different phase (solid, liquid, or gas) from the reactants; a contact catalyst.

Heterogeneous equilibria – equilibria involving species in more than one phase.

Heterogeneous mixture – a mixture that does not have uniform composition and properties throughout.

Heteronuclear – consisting of different elements.

High spin complex – crystal field designation for an outer orbital complex; t_{2g} and e_g orbitals are singly occupied before pairing occurs.

Homogeneous catalyst – in the same phase (solid, liquid, or gas) as the reactants.

Homogeneous equilibria – when *reagents* and *products* are of the same phase (i.e., gases, liquids, or solids).

Homogeneous mixture – a mixture which has uniform composition and properties throughout.

Homologous series – compounds with each member differing from the next by a specific number and kind of atoms.

Homonuclear – consisting of only one element.

Hund's rule – single electrons must occupy orbitals of a given sublevel before pairing begins; see *Aufbau* (or *building up*) *principle*.

Hybridization – mixing atomic orbitals to form a new set of atomic orbitals with the same electron capacity and properties and energies intermediate between the original unhybridized orbitals.

Hydrate – a solid compound that contains a definite percentage of bound water.

Hydrate isomers – crystalline complexes that differ in whether water exists inside or outside the coordination sphere.

Hydration – the reaction of a substance with water.

Hydration energy – the energy change accompanying the hydration of a mole of gas and ions.

Hydride – a binary compound of hydrogen.

Hydrocarbons – compounds that contain only carbon and hydrogen.

Hydrogen bond – a relatively strong dipole-dipole interaction (but still considerably weaker than the covalent or ionic bonds) between molecules containing hydrogen directly bonded to a small, highly electronegative atom, such as N, O or F.

Hydrogenation – the reaction in which hydrogen adds across a double or triple bond.

Hydrogen-oxygen fuel cell – hydrogen is the fuel (reducing agent) and oxygen is the oxidizing agent.

Hydrolysis – the reaction of a substance with water or its ions.

Hydrolysis constant – an equilibrium constant for a hydrolysis reaction.

Hydrometer – a device used to measure the densities of liquids and solutions.

Hydrophilic colloids – colloidal particles that repel water molecules.

I

Ideal gas – a hypothetical gas that obeys the postulates of the kinetic-molecular theory.

Ideal gas law – the product of pressure and the volume of an ideal gas is directly proportional to the number of moles of the gas and the absolute temperature. $PV = nRT$

Ideal solution – obeys Raoult's Law strictly.

Immiscible liquids – do not mix to form a solution (e.g., oil and water).

Indicators – for acid-base titrations, organic compounds that exhibit different colors in solutions of different acidities, determine the point at which the reaction between two solutes is complete.

Inert pair effect – characteristic of the post-transition minerals; the tendency of the electrons in the outermost atomic *s* orbital to remain un-ionized or unshared in compounds of post-transition metals.

Inhibitory catalyst – an inhibitor; a catalyst that decreases the rate of reaction.

Inner orbital complex – valence bond designation for a complex in which the metal ion utilizes d orbitals for one shell inside the outermost occupied shell in its hybridization.

Inorganic chemistry – a part of chemistry concerned with inorganic (non carbon-based) compounds.

Insulator – a material that resists the flow of electric current or heat transfer; does not allow heat to flow easily.

Insoluble compound – a substance that will not dissolve in a solvent, even after mixing.

Integrated rate equation – an expression giving the concentration of a reactant remaining after a specified time; has a different mathematical form for different orders of reactants.

Intermolecular forces – forces between individual particles (atoms, molecules, ions) of a substance.

Ion – a molecule that has gained or lost electrons; an atom or a group of atoms carries an electric charge (Na^+).

Ion product for water – equilibrium constant for water ionization; $K\text{w} = [H_3O^+]\cdot[^-OH] = 1.00 \times 10^{-14}$ at 25 °C.

Ionic bond – electrostatic attraction between oppositely charged ions, resulting from a transfer of electrons.

Ionic bonding – chemical bonding resulting from transferring electrons from one atom or group.

Ionic compounds – compounds containing predominantly ionic bonding.

Ionic geometry – arrangement of atoms (not lone pairs of electrons) about the central atom of a polyatomic ion.

Ionization – the breaking up of a compound into separate ions; in an aqueous solution, the process by which a molecular compound reacts with water and forms ions.

Ionization constant – equilibrium constant for the ionization of a weak electrolyte.

Ionization energy – the minimum amount of energy required to remove the most loosely held electron of an isolated gaseous atom or ion.

Ion exchange – a method of removing hardness from water, and it replaces the positive ions that cause the hardness with H^+ ions.

Ionization isomers – result from the interchange of ions inside and outside the coordination sphere.

Isoelectric – having the same electronic configurations.

Isomers – different substances with the same formula.

Isomorphous – refers to crystals having the same atomic arrangement.

Isotopes – two or more forms of atoms of the same element with different masses; atoms containing the same number of protons but different numbers of neutrons.

IUPAC – acronym for "International Union of Pure and Applied Chemistry."

J

Joule (J) – a unit of energy in the SI system; one joule is 1 $kg \cdot m^2/s^2$, which is 0.2390 calories.

K

K capture – absorption of a *K* shell (n = 1) electron by a proton as it is converted to a neutron.

Kelvin – a unit of measure for temperature based upon an absolute scale.

Kinetics – a sub-field of chemistry specializing in reaction rates.

Kinetic energy (*KE*) – energy that matter processes by its motion.

Kinetic-molecular theory – a theory that attempts to explain macroscopic observations on gases in microscopic or molecular terms.

L

Lanthanides – elements 57 (lanthanum) through 71 (lutetium); grouped because of their similar behavior in chemical reactions.

Lanthanide contraction – a decrease in the radii of the elements following the lanthanides compared to what would be expected if there were no *f*-transition metals.

Latent heat – the energy absorbed or released when a substance changes state without changing temperature.

Lattice – unique arrangement of atoms or molecules in a crystalline liquid or solid.

Law of combining volumes (Gay-Lussac's Law) – at constant temperature and pressure, the volumes of reacting gases (and any gaseous products) can be expressed as ratios of small whole numbers.

Law of conservation of energy – energy cannot be created or destroyed; it can only change form.

Law of conservation of matter – there is no detectable change in the quantity of matter during an ordinary chemical reaction.

Law of conservation of matter and energy – the amount of matter and energy in the universe is fixed.

Law of definite proportions (law of constant composition) – different samples of a pure compound contain the same elements in the same proportions by mass.

Law of partial pressures (or *Dalton's Law*) – the pressure exerted by a mixture of gases is the sum of the partial pressures of the individual gases.

Laws of thermodynamics – physical laws which define quantities of thermodynamic systems describe how they behave and (by extension) set certain limitations such as perpetual motion.

Lead storage battery – secondary voltaic cell used in most automobiles.

Leclanche cell – a common type of *dry cell.*

Le Chatelier's principle – states that a system at equilibrium, or striving to attain equilibrium, responds in such a way as to counteract any stress placed upon it; if stress (change of conditions) is applied to a system at equilibrium, the system will shift in the direction that reduces stress.

Leveling effect – acids stronger than the acid characteristic of the solvent reacts with the solvent produce that acid; a similar statement applies to bases. The strongest acid (base) that can exist in a given solvent is the acid (base) characteristic of the solvent.

Levorotatory – an optically active substance rotates plane-polarized light counterclockwise, known as a *levo*.

Lewis acid – any species that can accept a share in an electron pair.

Lewis base – any species that can make available a share in an electron pair.

Lewis dot formula (electron dot formula) – representation of a molecule, ion or formula unit by showing atomic symbols and only outer shell electrons.

Ligand – a Lewis base in a coordination compound.

Light – that portion of the electromagnetic spectrum visible to the naked eye; known as "visible light."

Limiting reactant – a substance that stoichiometrically limits the number of product(s) that can be formed.

Linear accelerator – a device used for accelerating charged particles along a straight line path.

Line spectrum – an atomic emission or absorption spectrum.

Linkage isomers – a particular ligand bonds to a metal ion through different donor atoms.

Liquid – a state of matter which takes the shape of its container.

Liquid aerosol – colloidal suspension of a liquid in gas.

London dispersion forces – very weak and very short-range attractive forces between short-lived temporary (induced) dipoles; known as "dispersion forces."

Lone pair – pair of electrons residing on one atom and not shared by other atoms; unshared pair.

Low spin complex – crystal field designation for an inner orbital complex; contains electrons paired t_{2g} orbitals before e_g orbitals are occupied in octahedral complexes.

Lubricant – a substance capable of reducing friction (i.e., force that opposes the direction of motion).

M

Magnetic field – a space around a magnet where magnetism can be detected.

Magnetic quantum number – quantum mechanical solution to a wave equation designating the orbital within a given set (*s*, *p*, *d*, *f*) in which an electron resides.

Manometer – a two-armed barometer.

Mass (*m*) – a measure of the amount of matter in an object; mass is usually measured in grams or kilograms.

Mass action expression – for a reversible reaction, aA + bB cC + dD; the product of the concentrations of the products (species on right), each raised to the power corresponding to its coefficient in the balanced chemical equation, divided by the product of the concentrations of reactants (species on left), each raised to the power corresponding to its coefficient in the balanced equation; at equilibrium the mass action expression is equal to *K*.

Mass deficiency – the amount of matter converted into energy when an atom forms from constituent particles.

Mass number – the sum of the numbers of protons and neutrons in an atom; an integer.

Mass spectrometer – an instrument that measures the charge-to-mass ratio of charged particles.

Matter – anything that has mass and occupies space.

Mechanism – the sequence of steps by which reactants are converted into products.

Melting point – the temperature at which liquid and solid coexist in equilibrium.

Meniscus – the shape assumed by the surface of a liquid in a cylindrical container.

Melting – the phase change from a solid to a liquid.

Metal – a chemical element that is a good conductor of electricity and heat and forms cations and ionic bonds with nonmetals; elements below and to the left of the stepwise division (metalloids) in the upper right corner of the periodic table; about 80% of known elements are metals.

Metallic bonding – bonding within metals due to the electrical attraction of positively charged metal ions for mobile electrons that belong to the crystal.

Metallic conduction – conduction of electrical current through a metal or along a metallic surface.

Metalloid – a substance with the properties of metals and nonmetals (B, Al, Si, Ge, As, Sb, Te, Po and At).

Metathesis reactions – reactions in which two compounds react to form two new compounds, with no changes in oxidation number; reactions in which the ions of two compounds exchange partners.

Method of initial rates – method of determining the rate-law expression by carrying out a reaction with different initial concentrations and analyzing the resultant changes in initial rates.

Methylene blue – a heterocyclic aromatic chemical compound with the molecular formula $C_{16}H_{18}N_3SCl$.

Miscible liquids – mix to form a solution (e.g., alcohol and water).

Miscibility – the ability of one liquid to mix with (dissolve in) another liquid.

Mixture – two or more different substances mingled together but not chemically combined. A sample of matter composed of two or more substances, each of which retains its identity and properties.

Moderator – a substance (e.g., deuterium, oxygen, paraffin) capable of slowing fast neutrons upon collision.

Molality – a concentration expressed as a number of moles of solute per kilogram of solvent.

Molarity – the number of moles of solute per liter of solution.

Molar solubility – the number of moles of a solute that dissolves to produce a liter of a saturated solution.

Mole – a measurement of an amount of substance; a single mole contains approximately 6.022×10^{23} units or entities; abbreviated mol.

Molecule – a chemically bonded number of electrically neutral atoms.

Molecular equation – a chemical reaction in which formulas are written as if substances existed as molecules; only complete formulas are used.

Molecular formula – indicates the actual number of atoms present in a molecule of a molecular substance.

Molecular geometry – the arrangement of atoms (not lone pairs of electrons) around a central atom of a molecule or polyatomic ion.

Molecular orbital (*MO*) – resulting from the overlap and mixing of atomic orbitals on different atoms (i.e., a region where an electron can be found in a molecule, as opposed to an atom); an MO belongs to the molecule.

Molecular orbital theory – a theory of chemical bonding based upon postulated molecular orbitals.

Molecular weight – the mass of one molecule of a nonionic substance in atomic mass units.

Molecule – the smallest particle of a compound capable of stable, independent existence.

Mole fraction – the number of moles of a component in a mixture divided by the number of moles in the mixture.

Monoprotic acid – can form only one hydronium ion per molecule; may be strong or weak.

Mother nuclide – nuclide that undergoes nuclear decay.

N

Native state – refers to the occurrence of an element in an uncombined or free state in nature.

Natural radioactivity – spontaneous decomposition of an atom.

Neat – conditions with a liquid reagent or gas performed with no added solvent or co-solvent.

Nernst equation – corrects standard electrode potentials for nonstandard conditions.

Net ionic equation – results from canceling spectator ions and eliminating brackets from a total ionic equation.

Neutralization – the reaction of an acid with a base to form a salt and water; usually, the reaction of hydrogen ions with hydrogen ions to form water molecules.

Neutrino – particle that can travel at speeds close to the speed of light; created due to radioactive decay.

Neutron – a neutral unit or subatomic particle with no net charge and a mass of 1.0087 amu.

Nickel-cadmium cell (NiCd battery) – a dry cell where the anode is Cd, the cathode is NiO_2, and the electrolyte is basic.

Nitrogen cycle – the complex series of reactions by which nitrogen is slowly but continually recycled in the atmosphere, lithosphere, and hydrosphere.

Noble gases – elements of the periodic Group 0; He, Ne, Ar, Kr, Xe, Rn; known as "rare gases;" formerly called "inert gases."

Nodal plane – a region in which the probability of finding an electron is zero.

Nonbonding orbital – a molecular orbital derived only from an atomic orbital of one atom; lends neither stability nor instability to a molecule or ion when populated with electrons.

Nonelectrolyte – a substance whose aqueous solutions do not conduct electricity.

Nonmetal – an element that is not metallic.

Nonpolar bond – a covalent bond in which electron density is symmetrically distributed.

Nuclear – of or about the atomic nucleus.

Nuclear binding energy – the energy equivalent of the mass deficiency; the energy released in forming an atom from the subatomic particles.

Nuclear fission – when a heavy nucleus splits into nuclei of intermediate masses and protons are emitted.

Nuclear magnetic resonance spectroscopy – a technique that exploits the magnetic properties of specific nuclei; helpful in identifying unknown compounds.

Nuclear reaction – involves a change in the composition of a nucleus and can emit or absorb a tremendous amount of energy.

Nuclear reactor – a system in which controlled nuclear fission reactions generate heat energy on a large scale, subsequently converted into electrical energy.

Nucleons – particles comprising the nucleus; protons, and neutrons.

Nucleus – the very small and dense, positively charged center of an atom containing protons and neutrons, as well as other subatomic particles; the net charge is positive.

Nuclides – refers to different atomic forms of elements; in contrast to isotopes, which refer only to different atomic forms of a single element.

Nuclide symbol – designation for an atom A/Z E, in which E is the symbol of an element, Z is its atomic number, and A is its mass number.

Number density – a measure of the concentration of countable objects (e.g., atoms, molecules) in space; the number per volume.

O

Octahedral – molecules and polyatomic ions with one atom in the center and six atoms at the corners of an octahedron.

Octane number – a number that indicates how smoothly a gasoline burns.

Octet rule – during bonding, atoms tend to reach an electron arrangement with eight electrons in the outermost shell. Many representative elements attain at least a share of eight electrons in their valence shells when they form molecular or ionic compounds; there are some limitations.

Open sextet – species with only six electrons in the highest energy level of the central element (many Lewis acids).

Orbital – may refer to an atomic orbital or a molecular orbital.

Organic chemistry – the chemistry of substances that contain carbon-hydrogen bonds.

Organic compound – substances that contain carbon.

Osmosis – when solvent molecules pass through a semi-permeable membrane from a dilute solution into a more concentrated solution.

Osmotic pressure – the hydrostatic pressure produced on the surface of a semi-permeable membrane.

Outer orbital complex – valence bond designation for a complex in which the metal ion utilizes d orbitals in the outermost (occupied) shell in hybridization.

Overlap – the interaction of orbitals on different atoms in the same region of space.

Oxidation – the addition of oxygen or the loss of electrons. An algebraic increase in the oxidation number; may correspond to a loss of electrons.

Oxidation numbers – quantitative values used as mechanical aids in writing formulas and balancing equations; for single-atom ions, they correspond to the charge on the ion; more electronegative atoms are assigned negative oxidation numbers, known as *oxidation states*.

Oxidation-reduction reactions – reactions in which oxidation and reduction occur; known as *redox reactions*.

Oxide – a binary compound of oxygen.

Oxidizing agent – the substance that oxidizes another substance and is reduced.

P

Pairing – a favorable interaction of two electrons with opposite m values in the same orbital.

Pairing energy – the energy required to pair two electrons in the same orbital.

Paramagnetism – attraction toward a magnetic field, stronger than diamagnetism but still weak compared to ferromagnetism.

Partial pressure – the force exerted by one gas in a mixture of gases.

Particulate matter – fine, divided solid particles suspended in polluted air.

Pauli exclusion principle – no electrons in the same atom may have identical sets of four quantum numbers.

Percentage ionization – the percentage of the weak electrolyte that will ionize in a solution of given concentration.

Percent by mass – 100% times the actual yield divided by the theoretical yield.

Percent composition – the mass percent of each element in a compound.

Percent purity – the percent of a specified compound or element in an impure sample.

Period – the elements in a horizontal row of the periodic table.

Periodicity – regular periodic variations of properties of elements with their atomic number (and position in the periodic table).

Periodic Law – the properties of the elements are periodic functions of their atomic numbers.

Periodic table – an arrangement of elements by increasing atomic numbers, emphasizes periodicity.

Peroxide – a compound with oxygen in –1 oxidation state; metal peroxides contain the peroxide ion, O_2^{2-}.

pH – the measure of acidity (or basicity) of a solution; negative logarithm of the concentration (mol/L) of the H_3O^+ $[H^+]$ ion; scale is commonly used over a range 0 to 14.

Phase diagram – shows the equilibrium temperature-pressure relationships for different phases of a substance.

pH scale – a range from 0 to 14. If the pH of a solution is 7 it is neutral; if the pH of a solution is less than 7 it is acidic; if the pH of a solution is greater than 7 it is basic.

Permanent hardness – hardness (relative to lathering soap) in water that cannot be removed by boiling; caused by calcium sulfate.

Photoelectric effect – emission of an electron from the surface of a metal caused by impinging electromagnetic radiation of specific minimum energy; the current increases with increasing radiation intensity.

Photon – a carrier of electromagnetic radiation of all wavelengths, such as gamma rays and radio waves; known as *quantum of light*.

Physical change – when a substance changes from one physical state to another, but no substances with different composition are formed; physical change may involve a phase change (e.g., melting, freezing) or another physical change such as crushing a crystal or separating one volume of liquid into different containers; does not produce a new substance.

Plasma – a physical state of matter that exists at extremely high temperatures in which molecules are dissociated, and most atoms are ionized.

Polar bond – a covalent bond with an unsymmetrical distribution of electron density.

Polarimeter – a device used to measure optical activity.

Polarization – the buildup of a product of oxidation or reduction of an electrode, preventing further reaction.

Polydentate – refers to ligands with more than one donor atom.

Polyene – a compound that contains more than one double bond per molecule.

Polymerization – the combination of many small molecules to form large molecules.

Polymer – a large molecule consisting of chains or rings of linked monomer units, usually characterized by high melting and boiling points.

Polymorphous – refers to substances that can crystallize in more than one crystalline arrangement.

Polyprotic acid – forms two or more hydronium ions per molecule; often, at least one ionization step is weak.

Positron – a nuclear particle with the mass of an electron but opposite charge (positive).

Potential difference (or *voltage*) – the force which moves the electrons around the circuit; the unit is Volt (V).

Potential energy (*PE*) – energy stored in a body or a system due to its position in a force field or configuration.

Power – the rate at which energy is converted from one form to another; the unit is Watts (W). Power = voltage × current ($P = VI$).

Precipitate – an insoluble solid formed by mixing in solution the constituent ions of a slightly soluble solution.

Precision – how close the results of multiple experimental trials are; see *accuracy*.

Pressure – force per unit area; unit is Pascal (Pa).

Primary standard – a known high degree of purity substance that undergoes one invariable reaction with the other reactant of interest.

Primary voltaic cells – voltaic cells that cannot be recharged; no further chemical reaction is possible once the reactants are consumed.

Products – chemicals produced (from reactants) in a chemical reaction.

Proton – a subatomic particle having a mass of 1.0073 amu and a charge of +1, found in the nuclei of atoms.

Protonation – the addition of a proton (H^+) to an atom, molecule or ion.

Pseudobinaryionic compounds – contain more than two elements but are named like binary compounds.

Q

Quanta – the minimum amount of energy emitted by radiation.

Quantum mechanics – the study of how atoms, molecules, subatomic particles, etc. behave and are structured; a mathematical method of treating particles based on quantum theory, which assumes that energy (of small particles) is not infinitely divisible.

Quantum numbers – numbers that describe the energies of electrons in atoms; derived from quantum mechanical treatment.

Quarks – elementary particles and a fundamental constituent of matter, combining to form hadrons (i.e., protons and neutrons).

R

Radiation – 1) heat transfer through invisible rays, which travel outwards from the hot object without a medium. 2) high-energy particles or rays emitted during the nuclear decay processes.

Radical – an atom or group of atoms that contains one or more unpaired electrons; usually a very reactive species.

Radioactive dating – method of dating ancient objects by determining the ratio of amounts of mother and daughter nuclides present in an object and relating the ratio to the object's age via half-life calculations.

Radioactive tracer – a small amount of radioisotope replacing a nonradioactive isotope of the element in a compound whose path (e.g., in the body) or whose decomposition products are monitored by detection of radioactivity; known as a "radioactive label."

Radioactivity – the spontaneous disintegration of atomic nuclei.

Raoult's Law – the vapor pressure of a solvent in an ideal solution decreases as its mole fraction decreases.

Rate-determining step – the slowest step in a mechanism; the step that determines the overall reaction rate.

Rate-law expression – equation relating the reaction rate to the concentrations of the reactants and the specific rate of the constant.

Rate of reaction – the change in the concentration of a reactant or product per unit time.

Reactants – substances consumed in a chemical reaction; react together in a chemical reaction.

Reaction quotient – the mass action expression under any set of conditions (not necessarily equilibrium); its magnitude relative to K determines the direction in which the reaction must occur to establish equilibrium.

Reaction ratio – the relative amounts of reactants and products involved in a reaction; may be the ratio of moles, millimoles, or masses.

Reaction stoichiometry – describes the quantitative relationships among substances participating in chemical reactions.

Reactivity series (or activity series) – an empirical, calculated, and structurally analytical progression of a series of metals, arranged by "reactivity" from highest to lowest; used to summarize information about the reactions of metals with acids and water, double displacement reactions and the extraction of metals from ores.

Reagent – a substance (or compound) added to a system to cause a chemical reaction or to visualize if a reaction occurs; the terms reactant and reagent are often used interchangeably; however, a *reactant* is more specifically a substance consumed during a chemical reaction.

Reducing agent – a substance that reduces another substance and is itself oxidized.

Reduction – the removal of oxygen or the gaining of electrons.

Resonance – the concept in which two or more equivalent dot formulas for the same arrangement of atoms (resonance structures) are necessary to describe the bonding in a molecule or ion.

Reverse osmosis – forcing solvent molecules to flow through a semi-permeable membrane from a concentrated solution into a dilute solution by applying greater hydrostatic pressure on the concentrated side than the osmotic pressure opposing it.

Reversible reaction – proceses that do not go to completion and occur in the forward and reverse direction.

S

Saline solution – a general term for NaCl (i.e., sodium chloride) in water.

Salt – when a metal replaces the hydrogen of an acid.

Salts – ionic compounds composed of anions and cations.

Salt bridge – a U-shaped tube containing an electrolyte, connects the two half-cells of a voltaic cell.

Saturated solution –no more solute will dissolve at that temperature.

***s*-block elements** – group 1 and 2 elements (alkali and alkaline metals), including hydrogen and helium.

Schrödinger equation – quantum state equation representing the behavior of an electron around an atom; describes the wave function of a physical system evolving.

Second Law of Thermodynamics – the universe tends toward a state of greater disorder in spontaneous processes.

Secondary standard – a solution that has been titrated against a primary standard; a standard solution.

Secondary voltaic cells – voltaic cells that can be recharged; original reactants can be regenerated by reversing the direction of the current flow.

Semiconductor – a substance that does not conduct electricity at low temperatures but will do so at higher temperatures.

Semi-permeable membrane – a thin partition between two solutions through which specific molecules can pass but others cannot.

Shielding effect – electrons in filled sets of *s*, *p* orbitals between the nucleus and outer shell electrons shield the outer shell electrons somewhat from the effect of protons in the nucleus; known as the "screening effect."

Sigma (σ) bonds – bonds resulting from the head-on overlap of atomic orbitals. The region of electron sharing is along and (cylindrically) symmetrical to the imaginary line connecting the bonded atoms.

Sigma orbital – molecular orbital resulting from the head-on overlap of two atomic orbitals.

Single bond – covalent bond resulting from the sharing of two electrons (one pair) between two atoms.

Sol – a suspension of solid particles in a liquid; artificial examples include sol-gels.

Solid – one of the states of matter, where the molecules are packed close, resistance to movement/deformation and volume change.

Solubility product constant – equilibrium constant that applies to the dissolution of a slightly soluble compound.

Solubility product principle – the solubility product constant expression for a slightly soluble compound is the product of the concentrations of the constituent ions, each raised to the power that corresponds to the number of ions in one formula unit.

Solute – the dispersed (i.e., dissolved) phase of a solution; the solution is mixed into the solvent (e.g., NaCl in saline water).

Solution – a homogeneous mixture of multiple substances; comprised of solutes and solvents; a mixture of a solute (usually a solid) and a solvent (usually a liquid).

Solvation – the process by which solvent molecules surround and interact with solute ions or molecules.

Solvent – the dispersing medium of a solution (e.g., H_2O in saline water).

Solvolysis – the reaction of a substance with the solvent in which it is dissolved.

***s*-orbital** – a spherically symmetrical atomic orbital; one per energy level.

Specific gravity – the ratio of the density of a substance to the density of water.

Specific heat – the amount of heat required to raise the temperature of one gram of substance one degree Celsius.

Specific rate constant – an experimentally determined (proportionality) constant, which is different for different reactions, and which changes only with temperature; *k* in the rate-law expression: Rate = k [A] × [B].

Spectator ions – ions in a solution that do not participate in a chemical reaction.

Spectral line – any of several lines corresponding to definite wavelengths of an atomic emission or absorption spectrum; marks the energy difference between two energy levels.

Spectrochemical series – arrangement of ligands in order of increasing ligand field strength.

Spectroscopy – the study of radiation and matter, such as X-ray absorption and emission spectroscopy.

Spectrum – display of component wavelengths (colors) of electromagnetic radiation.

Speed of light – the speed at which radiation travels through a vacuum (299,792,458 m/sec).

Square planar – describes molecules and polyatomic ions with one atom in the center and four atoms at the corners of a square.

Square planar complex – relationship with metal in the center of a square plane, with ligand donor atoms at each of the four corners.

Standard conditions for temperature and pressure (STP) – used to compare experimental results (25 °C and 100.000 kPa).

Standard electrodes – half-cells in which the oxidized and reduced forms of a species are present at the unit activity (1.0 M solutions of dissolved ions, 1.0 atm partial pressure of gases, pure solids, and liquids).

Standard electrode potential – by convention, the potential (Eo) of a half-reaction as a reduction relative to the standard hydrogen electrode when species are present at unit activity.

Standard entropy – the absolute entropy of a substance in its standard state at 298 K.

Standard molar enthalpy of formation – the amount of heat absorbed in forming one mole of a substance in a specified state from its elements in their standard states.

Standard molar volume – the space occupied by one mole of an ideal gas under standard conditions; 22.4 liters.

Standard reaction – a process where the numbers of moles of reactants in the balanced equation, in their standard states, are entirely converted to the numbers of moles of products in the balanced equation, at their standard state.

State of matter – a homogeneous, macroscopic phase (e.g., gas, plasma, liquid, solid) in increasing concentration.

Stoichiometry – the quantitative relationships among elements and compounds undergoing chemical changes.

Strong electrolyte – a substance that conducts electricity well in a dilute aqueous solution.

Strong field ligand – ligand that exerts a strong crystal or ligand electrical field and generally forms low spin complexes with metal ions when possible.

Structural isomers – compounds that contain the same number and kinds of atoms with different geometry.

Subatomic particles – comprise an atom (e.g., protons, neutrons, electrons).

Sublimation – the direct vaporization of a solid by heating without passing through the liquid state; a phase transition from solid to limewater fuel or gas.

Substance – any matter, specimens with the same chemical composition and physical properties.

Substitution reaction – a reaction in which another atom or group of atoms replaces an atom or a group of atoms.

Supercooled liquids – liquids that, when cooled, apparently solidify but continue to flow very slowly under the influence of gravity.

Supercritical fluid – a substance at a temperature above its critical temperature.

Supersaturated solution – contains a higher than saturation concentration of solute; slight disturbance or seeding causes crystallization of excess solute.

Suspension – a heterogeneous mixture in which solute-like particles settle out of the solvent-like phase sometime after their introduction. A mixture of a liquid and a finely divided insoluble solid.

T

Talc – a mineral representing the Mohs Scale composed of hydrated magnesium silicate with the chemical formula $H_2Mg_3(SiO_3)_4$ or $Mg_3Si_4O_{10}(OH)_2$.

Temperature – a measure of heat intensity (i.e., the hotness or coldness of a sample); a measure of the kinetic energy of an object. Unit is degrees, and scales are Celsius, Fahrenheit, and Kelvin.

Temporary hardness – hardness in water that can be removed by boiling; caused by calcium hydrogen carbonate.

Ternary acid – a ternary compound containing H, O and another element, often a nonmetal.

Ternary compound – a compound consisting of three elements; may be ionic or covalent.

Tetrahedral – a term used to describe molecules and polyatomic ions with one atom in the center and four atoms at the corners of a tetrahedron.

Theoretical yield – the maximum amount of a specified product that could be obtained from specified amounts of reactants, assuming complete consumption of the limiting reactant according to only one reaction and complete recovery of the product; see *actual yield.*

Theory – a model describing the nature of a phenomenon.

Thermal conductivity – a property of a material to conduct heat (often noted as k).

Thermal cracking – decomposition by heating a substance in the presence of a catalyst and the absence of air.

Thermochemistry – the study of absorption/release of heat within a chemical reaction; studies heat energy associated with chemical reactions and physical transformations.

Thermodynamics – studying the effects of changing temperature, volume, or pressure (or work, heat, and energy) on a macroscopic scale.

Thermodynamic stability – when a system is in its lowest energy state with its environment (equilibrium).

Thermometer – a device that measures the average energy of a system.

Thermonuclear energy – energy from nuclear fusion reactions.

Third Law of Thermodynamics – entropy of a pure crystalline substance at absolute zero temperature is zero.

Titration – a procedure in which one solution is added to another solution until the chemical reaction between the two solutes is complete; the concentration of one solution is known, and that of the other is unknown. The process of adding one solution to a measured amount of another to find out exactly how much of each is required to react.

Torr – a unit to measure pressure; 1 Torr is equivalent to 133.322 Pa or 1.3158×10^{-3} atm.

Total ionic equation – the expression for a chemical reaction written to show the predominant form of species in aqueous solution or contact with water.

Transition elements (metals) – B Group elements except IIB in the periodic table; sometimes called transition elements, elements with incomplete *d* sub-shells; the *d*-block elements.

Transition state theory –reactants pass through high-energy transition states before forming products.

Transuranic element – an atomic number greater than 92; none of the transuranic elements are stable.

Triple bond – the sharing of three pairs of electrons within a covalent bond (e.g., N_2).

Triple point – where the temperature and pressure of three phases are the same; water has a unique phase diagram.

Tyndall effect – results from light scattering by colloidal particles (a mixture where one substance is dispersed evenly throughout another) or by suspended particles.

U

Uncertainty –any measurement that involves estimating any amount that cannot be precisely reproducible.

Uncertainty principle – knowing the location of a particle makes the momentum uncertain, while knowing the momentum of a particle makes the location uncertain.

Unit cell – the smallest repeating unit of a lattice.

Unit factor – statements used in converting between units.

Universal (or ideal) gas constant – proportionality constant in the ideal gas law (0.08206 L·atm/(K·mol)).

UN number – a four-digit code used to note hazardous and flammable substances.

Unsaturated hydrocarbons – hydrocarbons that contain double or triple carbon-carbon bonds.

V

Valence bond theory – proposes that covalent bonds are formed when atomic orbitals on different atoms overlap and the electrons are shared.

Valence electrons – outermost electrons of atoms; usually those involved in bonding.

Valence shell electron pair repulsion theory (VSEPR) – assumes electron pairs are arranged around the central element of a molecule or polyatomic ion with maximum separation (and minimum repulsion) among regions of high electron density.

Valency – the number of electrons an atom wants to gain, lose, or share to have a full outer shell.

Van der Waals' equation – a quantitative relationship of a state extending the ideal gas law to real gases by including two empirically determined parameters, which are specific for different gases.

Van der Waals force – one of the forces (attraction/repulsion) between molecules.

Van't Hoff factor – the ratio of moles of particles in solution to moles of solute dissolved.

Vapor – when a substance is below the critical temperature in the gas phase.

Vaporization – the phase change from liquid to gas.

Vapor pressure – the particle pressure of vapor at the surface of its parent liquid.

Viscosity – the resistance of a liquid to flow (e.g., oil has a higher viscosity than water).

Volt – one joule of work per coulomb; the unit of electrical potential transferred.

Voltage – the potential difference between two electrodes; a measure of the chemical potential for a redox reaction.

Voltaic cells – electrochemical cells in which spontaneous chemical reactions produce electricity; known as *galvanic cells*.

Voltmeter – an instrument that measures the cell potential.

Volumetric analysis – measuring the volume of a solution (of known concentration) to determine the substance's concentration within the solution; see *titration*.

W

Water equivalent – the amount of water absorbing the same heat as the calorimeter per degree of temperature increases.

Weak electrolyte – a substance that conducts electricity poorly in a dilute aqueous solution.

Weak field ligand – a ligand exerting a weak crystal or ligand field and generally forms high spin complexes with metals.

X

X-ray – electromagnetic radiation between gamma and UV rays.

X-ray diffraction – a method for establishing structures of crystalline solids using single wavelength X-rays and studying the diffraction pattern.

X-ray photoelectron spectroscopy – a spectroscopic technique used to measure the composition of a material.

Y

Yield – the amount of product produced during a chemical reaction.

Z

Zone melting – remove impurities from an element by melting and slowly traveling it down an ingot (cast).

Zone refining – a method of purifying a metal bar by passing it through an induction heater; this causes impurities to move along a melted portion.

Zwitterion (formerly called a dipolar ion) – a neutral molecule with a positive and negative electrical charge; multiple positive and negative charges can be present, distinct from dipoles at different locations within that molecule; known as *inner salts*.

Essential Chemistry Self-Teaching Guides

Electronic Structure & Periodic Table

Chemical Bonding

States of Matter & Phase Equilibria

Stoichiometry

Solution Chemistry

Chemical Kinetics & Equilibrium

Acids & Bases

Chemical Thermodynamics

Electrochemistry

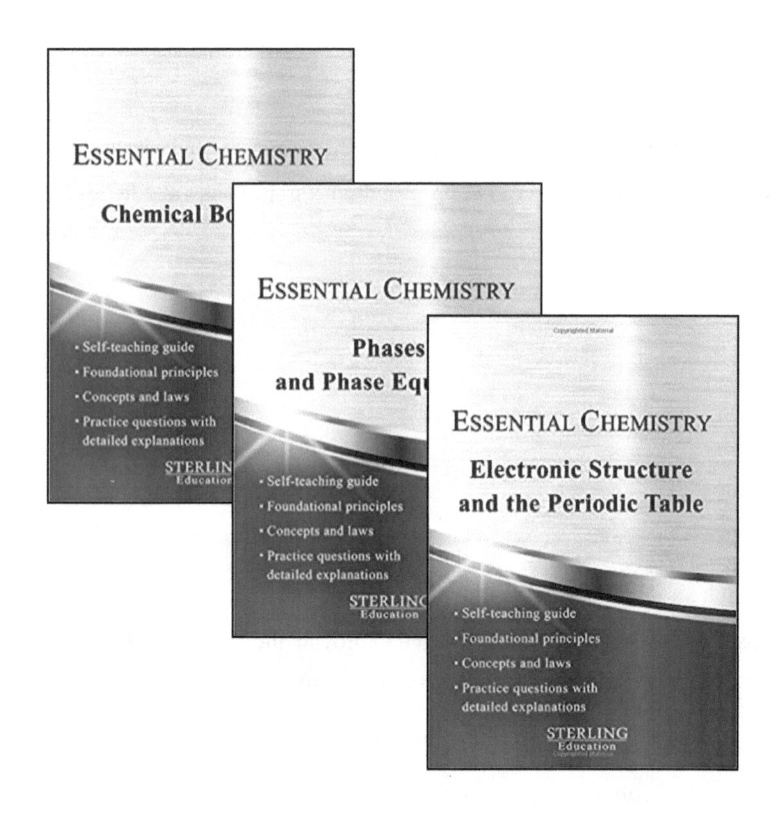

Everything You Always Wanted to Know About…

Chemistry

Physics

Cell and Molecular Biology

Organismal Biology

American History

American Law

American Government and Politics

Comparative Government and Politics

World History

European History

Psychology

Environmental Science

Human Geography

Visit our Amazon store

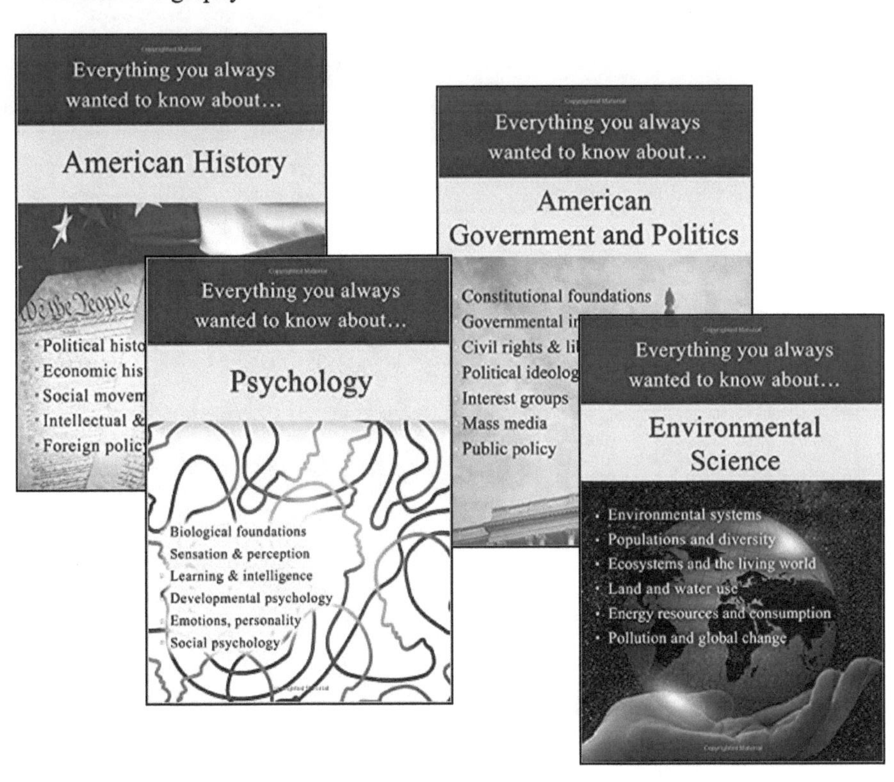

Made in United States
Troutdale, OR
05/07/2024